Lecture Notes in Mathematics 1681

Editors:
A. Dold, Heidelberg
F. Takens, Groningen

Springer
Berlin
Heidelberg
New York
Barcelona
Budapest
Hong Kong
London
Milan
Paris
Santa Clara
Singapore
Tokyo

Günther J. Wirsching

The Dynamical System Generated by the 3n+1 Function

 Springer

Author

Günther J. Wirsching
Katholische Universität Eichstätt
Mathematisch-Geographische Fakultät
D-85071 Eichstätt, Germany
e-mail: guenther.wirsching@ku-eichstaett.de

Cataloging-in-Publication Data applied for

Die Deutsche Bibliothek - CIP-Einheitsaufnahme

Wirsching, Günther:
The dynamical system generated by the 3n + 1 function / Günther J.
Wirsching. - Berlin ; Heidelberg ; New York ; Barcelona ; Budapest ;
Hong Kong ; London ; Milan ; Paris ; Santa Clara ; Singapore ;
Tokyo : Springer, 1998
  (Lecture notes in mathematics ; 1681)
  ISBN 3-540-63970-5

Mathematics Subject Classification (1991): 11B37, 60C05, 60B10, 11K41

ISSN 0075-8434
ISBN 3-540-63970-5 Springer-Verlag Berlin Heidelberg New York

© Springer-Verlag Berlin Heidelberg 1998
Printed in Germany

Typesetting: Camera-ready T_EX output by the author
SPIN: 10649733   46/3143-543210 - Printed on acid-free paper

THE DYNAMICAL SYSTEM
ON THE NATURAL NUMBERS
GENERATED BY THE $3n+1$ FUNCTION

Table of Contents

INTRODUCTION

Among the most fascinating mathematical problems are those which are easily formulated, but withstand for a long time sophisticated attacks for solving them. One instance of this kind of problems is the by now famous $3n + 1$ problem: Let $f(n) := n/2$, if n is even, and $f(n) := 3n + 1$, if n is odd. Choosing a natural number x as starting number and applying f repeatedly produces a sequence of natural numbers, which is called f-*trajectory* of x and denoted by

$$T_f(x) := \big(x, f(x), f(f(x)), \ldots, f^k(x), \ldots\big).$$

For example, taking $x = 13$ gives the f-trajectory

$$T_f(13) = (13, 40, 20, 10, 5, 16, 8, 4, 2, 1, 4, 2, 1, \ldots)$$

which continues periodically with the cycle $(4, 2, 1)$. All f-trajectories which have been calculated up to now have this limiting behaviour, and there are many starting numbers which have been tested (see section I.3). This leads to the $3n + 1$ *conjecture* which asserts that any f-trajectory eventually runs into the limiting cycle $(4, 2, 1)$. Is there a rigorous proof that this is the only possible limiting behaviour of a sequence of natural numbers generated by f?

Many authors consider the $3n + 1$ conjecture as intractably hard—and they may well be right, as the problem is still open. On the other hand, the problem is not new in the mathematical literature. Its—somewhat foggy—origin dates back to the 1930's; but since the 1970's we observe a rapidly growing interest in this problem and mathematics which people consider to be connected to it. (*Proof*: see the bibliography at the end of these notes.) If the $3n + 1$ conjecture itself appears to be intractable, what is, then, the mathematics people do around it, and is it really justified to claim that there is some progress towards a solution of the original problem? I do not plan to give an answer to this question in a few words, I just describe the facts and leave the final judgement to the reader.

The strategy for finding interesting things about an "intractable" problem is threefold: translate the conjecture into as many different contexts as you can, formulate weaker statements implied by the conjecture in question and try to prove some of them, and wait for flashes of genius giving new and interesting insights. In the case of the $3n+1$ problem, the conjecture has been reformulated, for instance, in terms of formal languages (see section I.12), and even in terms of analytic functions in the complex unit disk (see section I.13), leading to problems which seem as intractable as the original one. Following the second device, an "intermediate" conjecture is:

FINITE CYCLES CONJECTURE. *There are only finitely many cyclic numbers, i.e., the number of integers $y > 0$ such that $f^n(y) = y$ for some $n \in \mathbb{N}$ is finite.*

By now, this is also unproved. But there are some results in this direction: for example, R. P. Steiner proved in 1978, using a deep result of A. Baker on linear forms in logarithms, that there is just one cycle of a special type which had been called *circuit* (see section I.9).

A priori, an f-trajectory can either turn out to be eventually cyclic, or it must grow to infinity (this is due to the fact that f produces a *unique* successor to each number). Even the following consequence of the $3n + 1$ conjecture is unsolved.

(No) DIVERGENT TRAJECTORY CONJECTURE. *There is no divergent $3n + 1$ trajectory, i.e., there is no $y \in \mathbb{N}$ such that $\lim_{n \to \infty} f^n(y) = \infty$.*

But, also in this case there is a partial result: J. C. Lagarias showed in 1985 that, if a divergent f-trajectory happens to exist, then it cannot grow too slowly (see section I.6).

*   *   *

The point of view on the $3n + 1$ problem adopted here is based on a nice idea due to L. Collatz: he represented an arbitrary integer function $g : \mathbb{N} \to \mathbb{N}$, say, as a directed graph Γ_g with the domain $\mathbb{N}$ of g as infinite set of vertices, and with all pairs $(n, g(n))$ as directed edges. This graph derived from the function g is now called the *Collatz graph* of g. Taking $g := f$, with the integer function f defined above, we clearly have the following equivalence:

the $3n + 1$ conjecture holds $\Longleftrightarrow$ the graph Γ_f is (weakly) connected.

For a general integer function $g : \mathbb{N} \to \mathbb{N}$, the *(discrete) dynamical system* on $\mathbb{N}$ *generated* by g consists of all possible g-trajectories. Now it is clear that any g-trajectory must remain in some weak component of Γ_g. Moreover, two g-trajectories $\mathcal{T}_g(x)$ and $\mathcal{T}_g(y)$ *coalesce*, i.e. there are integers $n, m \geqslant 0$ such that $g^n(x) = g^m(y)$, if and only if x and y belong to the same weak component of the graph Γ_g. Taking a fixed g-trajectory $\mathcal{T}_g(x)$, the *domain of attraction* of this trajectory consists of all starting numbers $y \in \mathbb{N}$ whose g-trajectory coalesces with $\mathcal{T}_g(x)$. So we infer that a domain of attraction of the dynamical system on $\mathbb{N}$ generated by g is just a (weak) component of Γ_g. This means: the study of a dynamical system on $\mathbb{N}$ is equivalent to the study of a Collatz graph.

The topic of interest here is the dynamical system on $\mathbb{N}$ which is generated by the $3n + 1$ *function*

$$T : \mathbb{N} \to \mathbb{N}, \qquad T(n) := \begin{cases} T_0(n) := n/2 & \text{if } n \text{ is even,} \\ T_1(n) := (3n + 1)/2 & \text{if } n \text{ is odd.} \end{cases}$$

This function T replaces the function f defined above without loss of information: if n is even, then $T(n) = f(n)$, and if n is odd, then $T(n) = f(f(n))$ (as $3n + 1$

is even whenever n is odd). In this sense T "shortens" the f-trajectories; several authors prefer to deal with T instead of f. To study the dynamical system on $\mathbb{N}$ generated by T, we emphasize the *predecessor sets*

$$\mathcal{P}_T(a) := \{b \in \mathbb{N} : a \in \mathcal{T}_T(b)\} = \{b \in \mathbb{N} : \text{ some } T\text{-iterate of } b \text{ hits } a\}.$$

As any domain of attraction may be written as a union of predecessor sets, to study dynamical systems can mean to study predecessor sets. An interesting point about a predecessor set is any information refering to its *size*. In this number-theoretic setting, all information concerning the size of a set of natural numbers is contained in the *counting function* of that set. Here we consider counting functions of predecessor sets,

$$Z_a(x) := Z_{\mathcal{P}_T(a)}(x) := \big|\{n \in \mathcal{P}_T(a) : n \leqslant x\}\big|.$$

For technical reasons, it is easier to deal with predecessor sets of *non-cyclic* numbers, i.e. to natural numbers which do not pertain to a T-cylce. This is not really a restriction, as any domain of attraction can be written as a disjoint union of a trajectory and some predecessor sets of non-cyclic numbers (which can be chosen pairwise disjoint). For example, the domain of attraction of the T-cycle $(1, 2)$ is given by $\{1, 2\} \cup \mathcal{P}_T(4)$ (observe that $a \in \mathcal{P}_T(a)$ for each $a \in \mathbb{N}$). At this stage, we are still very close to the $3n + 1$ conjecture, as there are the equivalences:

$$\text{the } 3n + 1 \text{ conjecture holds} \quad \Longleftrightarrow \quad \mathcal{P}_T(4) = \mathbb{N} \setminus \{1, 2\}$$
$$\Longleftrightarrow \quad Z_4(x) = x - 2 \quad \text{for integers } x \geqslant 2.$$

But the $3n + 1$ conjecture itself may be intractable. So, we have to look for less ambitious assertions for treatment. Let us first state some properties which a general dynamical system on $\mathbb{N}$ given by $g : \mathbb{N} \to \mathbb{N}$ may or may not have.

POSITIVE PREDECESSOR DENSITY PROPERTY FOR FIXED $a \in \mathbb{N}$:

$$\liminf_{x \to \infty} \frac{Z_a(x)}{x} > 0 \,.$$

UNIFORM POSITIVE PREDECESSOR DENSITY ON $A \subset \mathbb{N}$:

$$\liminf_{x \to \infty} \left(\inf_{a \in A} \frac{Z_a(ax)}{x} \right) > 0 \,.$$

Note that the $3n + 1$ conjecture implies that the $3n + 1$ function shares the positive predecessor density property for $a = 4$. Uniform positive density is only interesting for infinite sets A; it seems reasonable to take $A := \{a \in \mathbb{N} : a \not\equiv 0 \mod 3\}$, as the predecessor sets of multiples of 3 are easily calculated:

$$\mathcal{P}_T(3k) = \{2^n \, 3k : n \in \mathbb{N}_0\}, \qquad \text{hence} \qquad Z_{3k}(x) = \left\lfloor \log_2 \frac{x}{3k} \right\rfloor.$$

Whether or not the $3n + 1$ function has the uniform positive density property seems to be quite independent from the $3n + 1$ conjecture.

Neither of these two properties is known for the $3n + 1$ function. But there has been some progress in a much weaker formulation of the uniform positive predecessor density property.

FIND GOOD EXPONENTS $c > 0$ SUCH THAT

$$\liminf_{x \to \infty} \left(\inf_{a \not\equiv 0 \bmod 3} \frac{Z_a(ax)}{x^c} \right) > 0 .$$

There has been a certain industry in improving the exponent c. The first who established that there is such an exponent $c > 0$ was R. E. Crandall (1978); actually Crandall derived his estimate only for the counting function $Z_1(x)$, but we shall see in section II.6 that his method also proves the relation above. Crandall's method has been pushed further by J. W. Sander (1987) to give $c = \frac{1}{4}$ (who also formulated it only for $Z_1(x)$). Finally, D. Applegate and J. C. Lagarias (1995) called Crandall's approach *tree-search method* and improved it to produce a computer-aided proof for $c = 0.654$. The tree-search method is related to the approach given here. In section II.6, we discuss tree-search in our terminology established in chapter II, showing that, in fact, the above uniform lim inf-relation is what has been proved.

The best result in this direction known up to now is the theorem of Applegate and Lagarias (1995) stating that, for each $a \not\equiv 0 \bmod 3$, there is a constant $c_a > 0$ such that $Z_a(x) \geq c_a\, x^{0.81}$ for each $x \geq a$. The computer-assisted proof is based on a different idea called *Krasikov inequalities* and initiated by I. Krasikov (1989). Although Krasikov inequalities appear more powerful in improving the constant c, this method is not discussed in these notes because it is not directly related to the approach prosecuted here.

One of the main results of these notes is the reduction theorem linking distribution properties of sums of mixed powers to dynamic properties like those stated above. To be more explicit about this, let j, k denote two non-negative integers. Then we are concerned with sums of mixed powers

$$2^{\alpha_0} + 2^{\alpha_1}3 + 2^{\alpha_2}3^2 + \cdots + 2^{\alpha_j}3^j ,$$

where $j + k \geq \alpha_0 > \cdots > \alpha_j \geq 0$. The set of all such sums will be denoted by $\mathcal{R}_{j,k}$. Then the cardinality of this set will be proved to be just the number of possible choices of integers $\alpha_0, \ldots, \alpha_j$ satisfying the condition above. This is elementary combinatorics:

$$|\mathcal{R}_{j,k}| = \binom{j+k+1}{j+1} .$$

Now the question is: given an integer $\ell \geq 1$, for which indices j, k does the set $\mathcal{R}_{j,k}$ meet all prime residue classes to modulus 3^ℓ (observe that an element of $\mathcal{R}_{j,k}$ cannot be divisible by 3, hence $\mathcal{R}_{j,k}$ is contained in the union of the prime residue classes to modulus 3^ℓ)? Technically, we do not want to deal with two indices j, k, but we want to deal with large sets $\mathcal{R}_{j,k}$. Hence, let us restrict attention to the sets $\mathcal{R}_{j-1,j}$ where the binomial coefficient is $\binom{2j}{j}$. The unsolved problem is the following.

COVERING CONJECTURE FOR MIXED POWER SUMS. *There is a constant $K > 0$ such that, for every $j, \ell \in \mathbb{N}$, the following implication holds:*

$$|\mathcal{R}_{j-1,j}| \geqslant K \cdot 2 \cdot 3^{\ell-1}$$

$$\implies \quad \mathcal{R}_{j-1,j} \text{ covers the prime residue classes to modulus } 3^\ell.$$

This conjecture seems reasonable: as there are precisely $2 \cdot 3^{\ell-1}$ prime residue classes modulo 3^ℓ, the precondition says that the set $\mathcal{R}_{j-1,j}$ has sufficient elements to put at least K of them into each prime residue class modulo 3^ℓ. If the distribution of $\mathcal{R}_{j-1,j}$ among those prime residue classes is not too unbalanced, one should expect that, for large K, we find at least one of the mixed power sums of $\mathcal{R}_{j-1,j}$ in each prime residue class. Of course, the essential content of the covering conjecture is in the asymptotics $\ell \to \infty$. If the conjecture is true, then it ensures that there is a sequence $(j_\ell)_{\ell \in \mathbb{N}}$ satisfying the two conditions:

(i) each set $\mathcal{R}_{j_\ell - 1, j_\ell}$ covers the prime residue classes modulo 3^ℓ, and

(ii) $\displaystyle \lim_{\ell \to \infty} \frac{j_\ell}{\ell} = \log_4 3$.

These two conditions will be technically essential in the proof of the following reduction theorem: *If the covering conjecture for mixed power sums is true,* then the dynamical system on $\mathbb{N}$ generated by the $3n + 1$ function has the following

UNIFORM SUB-POSITIVE PREDECESSOR DENSITY PROPERTY:

$$\liminf_{x \to \infty} \left(\inf_{a \not\equiv 0 \bmod 3} \frac{Z_a(ax)}{x^\delta} \right) > 0 \qquad \text{for any } \delta \in \mathbb{R} \text{ satisfying } 0 < \delta < 1.$$

In fact, the implication remains valid if the covering conjecture for mixed power sums is slightly weakened. We need not assume that there is a *constant* $K > 0$ with the required property. It suffices to assume that K is a function of ℓ which grows *sufficiently slowly* when ℓ tends to infinity; more explicitly, it suffices to assume that $K(\ell) e^{-\gamma \ell}$ remains bounded for any constant $\gamma > 0$.

The proof given here for the reduction of the uniform sub-positive predecessor density property to a covering conjectures for mixed power sums requires asymptotic analysis of binomial coefficients. In addition, we make use of integration theory on the compact topological group $\mathbb{Z}_3^*$ of invertible 3-adic integers. The proof gives, in addition, some argument why we cannot prove a uniform *positive* predecessor density property on the basis of a covering conjecture for mixed power sums like that given above. This has to do with the fact that the precise asymptotics of the binomial coefficient $\binom{2j}{j}$ is less than a constant times 2^{2j}. So the feeling arises that the $3n + 1$ predecessor sets may not have the uniform positive density property. In this context, we also give a technical condition on the distribution of mixed power sums which is sufficient for the positive predecessor density property for an individual $a \in \mathbb{N}$ (not divisible by 3, of course).

The intuitive content of the reduction is the following. The sums of mixed powers described above can be seen as the "accumulated non-linearities" occuring when iterating the $3n + 1$ function T. Then the conjecture on the distribution of power-sums states that the accumulated non-linearities of the $3n + 1$ function behave chaoticly. On the other hand, the uniform sub-positive predecessor density property would mean some regular behaviour of the dynamical system generated by T. So the reduction theorem means, intuitively, the more chaotic the behaviour of the accumulated non-linearities, the more regular is the behaviour of the dynamical system.

There are also many other results about the mathematical nature of the Collatz graph of T, or, equivalently, about the dynamical system generated by T—which justifies the title. Let us first briefly describe what is basic for collecting mathematical information about this dynamical system. A dynamical system consists of its trajectories. As there are difficulties in describing the limiting behaviour of the trajectories, we are forced to restrict our attention, at least at the beginning, on finite portions of trajectories. A finite portion of a T-trajectory, from b to a, say, has the form

$$b \xrightarrow{T} T(b) \xrightarrow{T} T^2(b) \xrightarrow{T} \ldots \xrightarrow{T} a = T^{k+\ell}(b).$$

Let us assume that on the $k+\ell$ steps from b to a, the function T takes precisely k times the branch T_0 and precisely ℓ times the branch T_1. The number of possible such (k, ℓ)-step portions of T-trajectories terminating at $a \in \mathbb{N}$ will be denoted by (observe that, if k and ℓ are fixed, then there is a one-one-correspondence between the (k, ℓ)-step portions and their initial vertices b, even if a is an element of a T-cycle)

$$e_\ell(k, a) := \left|\left\{ b \in \mathbb{N} : T^{k+\ell}(b) = a, k \text{ times } T_0, \ell \text{ times } T_1. \right\}\right|.$$

These quantities $e_\ell(k, a)$ constitute the basic objects for most of the research presented in this book. They are linked to $3n + 1$ predecessor density estimates via the implication, which is valid for non-cyclic numbers $a \in \mathbb{N}$ (theorem III.2.5),

$$\liminf_{n \to \infty} \frac{s_n(a)}{\beta^n} > 0$$
$$\implies \quad Z_a(x) \geqslant C \left(\frac{x}{a}\right)^{\log_2 \beta} \text{ for some constant } C > 0 \text{ and large } x,$$

where $s_n(a)$ is the n-th estimating series

$$s_n(a) := \sum_{\ell=1}^{\infty} e_\ell \left(n + \left\lfloor \ell \log_2 \frac{3}{2} \right\rfloor, a \right).$$

An important observation is that $e_\ell(k, a)$ depends on a only through its residue class to modulus 3^ℓ (this is one reason why we use ℓ as an index), which implies that we are concerned with a family of functions

$$e_\ell(k, \cdot) : \mathbb{Z}_3 \to \mathbb{N}_0 \quad \text{where } k, \ell \text{ run through } \mathbb{N}_0;$$

here $\mathbb{Z}_3$ denotes the group of 3-adic integers. A simple consideration shows that $e_\ell(k, a) = 0$ whenever $\ell \geqslant 1$ and $3 \mid a$. Hence, the set $\{e_\ell(k, \cdot) \mid \ell \geqslant 1, k \geqslant 0\}$ is a family of functions on the compact topological group $\mathbb{Z}_3^*$ of invertible 3-adic integers. The use of 3-adic integers in the context of the $3n + 1$ problem first appeared in [**Wir3**] (1994); the group of invertible 3-adic integers has also been connected to $3n + 1$ iterations, in a somewhat different setting, by Applegate and Lagarias [**AL3**] (1995).

As the domain of definition $\mathbb{Z}_3^*$ of our basic functions $e_\ell(k, \cdot)$ is a compact topological group, it admits a unique normalized Haar measure. It turns out that the $e_\ell(k, \cdot)$ are integrable w.r.t. this Haar measure, with the 3-*adic average*

$$\overline{e}_\ell(k) := \int_{\mathbb{Z}_3^*} e_\ell(k, a)\, da = \frac{1}{2 \cdot 3^{\ell-1}} \binom{k + \ell}{\ell}.$$

We obtain, for instance, to the following results:

(1) The estimating series given above give rise to a sequence of functions $s_n : \mathbb{Z}_3^* \to \overline{\mathbb{N}}_0$ which turns out to be discontinuous (theorem III.2.7) but perfectly Haar integrable (lemma III.3.6).

(2) The following is true (theorem III.5.2):

$$\liminf_{n \to \infty} \frac{1}{2^n} \int_{\mathbb{Z}_3^*} s_n(a)\, da > 0.$$

This means: If a number $a \in \mathbb{N}$ with $a \not\equiv 0 \bmod 3$ happens to have the property $e_\ell(k, a) = \overline{e}_\ell(k)$ for an appropriate portion of the pairs (k, ℓ), then the predecessor set $\mathcal{P}_T(a)$ has positive asymptotic density. Of course, the conclusion remains valid if, for an appropriate portion of the pairs (k, ℓ), $e_\ell(k, a)$ is sufficiently close to the 3-adic average.

(3) The numbers $e_\ell(k, a)$ can be constructed inductively without reference to the Collatz graph (corollary II.4.4). After an—admittedly complicated—normalization procedure (section IV.2), it turns out that the essential ingredient is an *asymptotically homogeneous* Markov chain in the sense that sequences of transition measures converge vaguely (theorem IV.4.1). Moreover, the limiting transition probability is averaging over $a \in \mathbb{Z}_3^*$.

(4) The limiting transition probability is shown to admit exactly one invariant density (theorem IV.5.1), which comes from a $\mathcal{C}^\infty$ function on $\mathbb{R}$ which is a polynomial on each interval outside the classical Cantor set (lemma IV.5.3).

The techniques to prove result (2) are essential for the proof of the reduction theorem discussed above. Result (3) embodies a first vague idea of "asymptotic self-similarity" of the Collatz graph. It would be nice to know more about this asymptotic self-similarity, and to compare it to the phenomena occurring in the context of discrete-time dynamical systems in the complex plane.

In these notes, I almost everywhere resisted the temptation to generalize a result to other functions than the $3n + 1$ function. A natural candidate for such generalizations would be a function T_q defined like the $3n + 1$ function T, but with $T_q(n) := (qn + 1)/2$ for odd n, where q is a previously fixed odd natural number. If q has the property that 2 generates the multiplicative group of prime residue classes to modulus q^ℓ for each $\ell \in \mathbb{N}$, then a good part of the results presented here admit a straightforward generalization to apply to iterations of T_q.

*   *   *

The plan of the book is as follows. Chapter I gives a brief survey of some strains of research on the problem. We already find a broad variety of different mathematical methods which have been used to attack the $3n + 1$ problem, including probability analysis, continued fractions, formal languages, and holomorphic functions in the complex unit disc.

In Chapter II, the essential notions for discrete dynamical systems on $\mathbb{N}$ and for the Collatz graph are given. Then the counting functions $e_\ell(k, a)$ and some variations thereof are introduced and discussed. These counting functions are linked to lower estimates for the predecessor counting functions $Z_a(x)$ via a series similar to, and, in fact, the model for the estimating series described above. It is shown that known density estimates (e.g., by Crandall [**Cra**] (1978) or Applegate and Lagarias [**AL1**] (1995)) for $3n + 1$ predecessor sets perfectly fit into our framework, which leads to slightly stronger formulation of those estimates.

Chapter III mainly deals with the use of 3-adic numbers. Motivated by the estimate for $Z_a(x)$ given in chapter I, the estimating series are introduced and discussed. The remaining part of chapter III studies their 3-adic averages, coming across a rigorous counterpart of the usual $3n + 1$ heuristics saying that in an "average" finite portion of a $3n + 1$ trajectory, the steps arising from T_0 and those arising from T_1 are, more or less, balanced. The chapter concludes with a proof of result (3) mentioned above, and with a short discussion of possible generalizations to $pn + 1$ functions for so-called Wieferich primes p.

Chapter IV contains all the stuff pertaining to the asymptotically homogeneous Markov chain mentioned above. The construction of this Markov chain is given explicitly, including descriptions of the transition probabilities in terms of combinatorial number theory. These explicit descriptions are used to formulate and prove the result indicated under (2) above.

Finally, chapter V takes up essential ideas from the previous chapters to prove the reduction theorem.

It is a great pleasure for me to include a long list of people who stimulated and supported my research on the $3n + 1$ problem in one way or another. First of all, let me express special gratitude to K. P. Hadeler, J. C. Lagarias, K. R. Matthews, G. Meinardus, D. Merlini, and H. A. Müller, for giving me access to important information around the $3n + 1$ problem, which I hardly could have received otherwise. I am indebted to L. Berg and J. C. Lagarias for valuable comments on a preliminary version of this. For further stimulating information, discussions on the topic, or encouragement to proceed, I also thank M. Kudlek, J. W. Sander, P. Schorn, and the following (in part former) mathematicians at the university of Eichstätt: A. Cornea, S. Deschauer, V. Krafft, J. Rohlfs, H.-G. Weigand, R. Wittmann, and especially W. Rump who originally drew my attention to the problem. Finally, I am grateful to those people who answered my request for sending reprints or preprints of their papers on the problem, and who are not yet mentioned: J. Błazewicz, S. Eliahou, G.-G. Gao, I. Korec, G. T. Leavens, and B. Schuppar.

Eichstätt, october 1997

G. J. Wirsching

CHAPTER I

SOME IDEAS AROUND $3n + 1$ ITERATIONS

The $3n+1$ problem can be found in many places. It is presented in D. R. Hofstadter's well-known book *Gödel, Escher, Bach* [**Hof**] (1980), pp. 400–402, where a natural number satisfying the $3n + 1$ conjecture (see section 1 for a precise statement) is called a *wondrous* number. The problem has been described in M. Gardner's article [**Grd**] (1972) in *Scientific American* and in C. S. Ogilvy's book *Tomorrow's math* [**Ogi**] (1972), p. 103f. It found entrance in R. K. Guy's problem book [**Guy1**] (1981); Guy also wrote some further introductory articles about $3n + 1$ iterations [**Guy2**] (1983), [**Guy3**] (1986).

In addition, there are more than fifty research articles containing substantial results around the $3n + 1$ problem. This chapter includes hints to some of the most important strains of research about this topic; thereby we come across a wealth of different mathematical ideas. The material is organized roughly according to themes, and inside a special topic according to date of publication.

1. The problem

The set of natural numbers (starting with 1) is denoted by $\mathbb{N} = \{1, 2, 3, \dots\}$. If we want to include 0, we write $\mathbb{N}_0 = \{0\} \cup \mathbb{N}$. The set of integers is denoted by $\mathbb{Z} = -\mathbb{N} \cup \mathbb{N}_0$. For an arbitrary integer function $f : \mathbb{Z} \to \mathbb{Z}$, we denote by $f^k = f \circ f^{k-1}$ the k-fold iterate of f, for each $k \in \mathbb{N}$, with the (natural) convention $f^0 = \mathrm{id}$. If $n \in \mathbb{Z}$, the f-*trajectory* of n (or with *starting number* n) is the sequence

$$\mathcal{T}_f(n) := \left(f^k(n)\right)_{k \geqslant 0} = \left(n, f(n), f \circ f(n), f \circ f \circ f(n), \dots\right).$$

An f-*cycle* is generated by an integer a with the property $f^k(a) = a$ for some $k \in \mathbb{N}$. For notational definiteness, we choose the minimal period k and write a cycle as a k-vector,

$$\Omega_f(a) := \left(a, f(a), \dots, f^{k-1}(a)\right).$$

In the german retroversion of [**Col2**] (1986),* L. Collatz calls the function

$$(1.1) \qquad\qquad f(n) = \begin{cases} 3n + 1 & \text{für ungerade } n \\ \frac{n}{2} & \text{für gerade } n, \end{cases}$$

---

*This paper originally was written in german, and then translated into chinese by Ren Zhiping; only the chinese translation is published. I am grateful to G. Meinardus, who sent me both the chinese version, and a retranslation into german by Zhangzheng Yu (1991).

the "$3n + 1$"-*Funktion*. Meanwhile it turned out to be more convenient to use instead the function

$$(1.2) \qquad T : \mathbb{N} \to \mathbb{N}, \qquad T(n) = \begin{cases} n/2 & \text{if } n \text{ is even,} \\ (3n + 1)/2 & \text{if } n \text{ is odd,} \end{cases}$$

(cf., for instance, [**Ter1**], [**Lag1**], [**BeM**], and most of the articles cited in our bibliography. Henceforth in these notes, this function T will be called the $3n + 1$ *function*, and we shall refer to that function f as the *Collatz function*. The letter T will be reserved for the $3n + 1$ function, but we do not reserve the letter f for the Collatz function (even in [**Col2**], f is also used to denote other functions). In some papers, T is called $3x + 1$ function, but I prefer the name $3n + 1$ function to emphasize that the problem is to deal with *natural* numbers.

The famous problem about the $3n + 1$ function is the following

$3n + 1$ CONJECTURE. *For any starting number in* $\mathbb{N}$, *the* T-*trajectory eventually ends in the cycle* $(1, 2)$.

A fallacious "proof" of this conjecture has been published by M. Yamada [**Yam**] (1980). The error has been described by J. C. Lagarias in his review (see also [**Yam**] for a citation). Prizes have been offered for its solution: \$ 50 by H. S. M. Coxeter in 1970, then \$ 500 by Paul Erdős , and £ 1000 by B. Thwaites [**Lag1**].

2. About the origin of the problem

The exact date of the first occurence of the $3n + 1$ conjecture is unclear. L. Collatz reports in [**Col2**] (1986) that he represented integer functions by graphs (for the precise definition, see chapter II) already in his student days from 1928 to 1933. He considered a certain classification of the possible graphs and tried to find simple examples for each type. Looking for a graph containing a "Kreis" (which is a cycle in our terminology) and representing a function f which should be as simple as possible, he was led to the observation that necessarily $f(n) < n$ for certain numbers n, and $f(n) > n$ for others. The first attempt was

$$(2.1) \qquad \widehat{f}(n) := \begin{cases} \frac{n}{2} & \text{if } n \text{ is even,} \\ n + 1 & \text{if } n \text{ is odd,} \end{cases}$$

which gives only the cycle $(1, 2)$, as is easily shown. The second attempt $\widehat{f}(n) := 2n + 1$ for odd n does not give any cycle at all, as odd numbers are mapped to larger odd numbers. And the next attempt is the Collatz function (1.1), of which Collatz reports that the only cycle he found was "der triviale Kreis" $(4, 2, 1)$. He writes that he did not publish the problem because he was unable to solve it.

Collatz also reports that he told the problem to his colleague H. Hasse in 1952. Hasse apparently circulated it by mouth during a visit to Syracuse university in the 1950's, where he proposed the name *Syracuse problem*. Later on, the problem also received the names *Kakutani's problem* and *Ulam's problem*, see [**Lag1**].

Independently of all that, B. Thwaites discovered the $3n+1$ problem at 4 p.m. on Monday, July 21st, 1952, and called it *Thwaites' conjecture*, abbreviated TC [**Thw**] (1985).

3. Empirical investigations and stochastic models

The $3n + 1$ literature contains several descriptions of empirical investigations. For instance, there is section devoted to this topic in the book by J. Nievergelt et al. [**NFR**] (1974), and we find a chapter on the $3n + 1$ problem in M. Jeger's introductory book [**Jeg**] (1986). Numerical algorithms for verifying the $3n + 1$ conjecture are considered by J. Arsac [**Ars**] (1987) and by H. Glaser and H.-G. Weigand [**GlW**] (1989). C. A. Pickover [**Pic**] (1989) exhibits graphical representations of numerical calculations of $3n + 1$ iterations.

Empirical investigations have reached a climax in the papers of G. T. Leavens [**Lea**] (1989) and G. T. Leavens and M. Vermeulen [**LeV**] (1992). They used sophisticated algorithms to determine *peaks* of certain functions $\phi : \mathbb{N} \to \mathbb{N}_0 \cup \{\infty\}$ related to the $3n + 1$ problem. Here a *peak* of ϕ is a natural number m such that $1 \leqslant n < m$ implies $\phi(n) < \phi(m)$. Leavens and Vermeulen use both the Collatz function f of (1.1), which they call H for *Hailstone function*, and the $3n + 1$ function T of (1.2). Among the functions from which they compute peaks are

$$\text{max\_value}(n) := \sup \left\{ k \in \mathbb{N}_0 : f^k(n) \right\} ,$$

the *maximal value* occuring in the f-trajectory of n,

$$\sigma(n) := \begin{cases} \infty & \text{if } T^k(n) \geqslant n \text{ for each } k \in \mathbb{N}, \\ \min\{k \in \mathbb{N} : T^k(n) < n\} & \text{otherwise,} \end{cases}$$

which is called *stopping time* of n (cf. section 6), and

$$\text{steps}(n) := \begin{cases} \infty & \text{if } f^k(n) \neq 1 \text{ for each } k \in \mathbb{N}, \\ \min\{k \in \mathbb{N} : f^k(n) = 1\} & \text{otherwise,} \end{cases}$$

which counts the number of iteration steps of the Collatz function needed to reach 1, when starting at n. In [**LeV**], there are tables of peaks of various functions, including max_value, σ, and steps, with data of size up to more than 5×10^{13}. The most remarkable numbers below 1000 are 27 and 703, which are peaks w.r.t. each of the three functions above; their values are

$$\text{max\_value}(27) = 9\,232, \qquad \sigma(27) = 59, \qquad \text{steps}(27) = 111,$$
$$\text{max\_value}(703) = 250\,504, \qquad \sigma(703) = 81, \qquad \text{steps}(703) = 130.$$

Heuristic reasoning with stochastic flavour appears frequently in publications touching upon the $3n+1$ problem; see, for instance, [**Hay**] (1984), [**Raw**] (1985), [**Wag**] (1985). The $3n + 1$ problem also has been presented in the context of statistical physics [**AMT**] (1989), [**FMR**] (1994).

It is tempting to enlarge empirical investigations by the use of stochastic processes designed to model $3n + 1$ iterations. Such a stochastic model would allow to calculate expected values of certain functions related to $3n+1$ iterations, and it is nice to compare such predictions to empirical data. Stochastic models are carefully analysed by J. C. Lagarias and A. Weiss [**LaW**] (1992); they find that certain predictions coming from two stochastic models agree with empirical data from $3n + 1$ iterations starting with $n \leqslant 10^{11}$. In this context, more empirical investigations have been carried out by D. Merlini et al. [**FMMT**] (1995), [**MSS1**] (1995), [**MSS2**] (1996).

Very interesting in this context is a recent paper of D. Applegate and J. C. Lagarias [**AL3**] (1995). Here *computed* data for $3n + 1$ predecessor trees are compared to *predicted* data derived from a branching process designed to model the growth of such predecessor trees. The result is astonishing: the range of variation of the computed data appears significantly narrower than that of the branching process models.

The value of stochastic models is not exhausted in a comparison of predictions and empirical data. Indeed, a stochastic model may have considerable heuristic value on a path to rigorous results. For instance, the use of Chernoff's "large deviation" bounds in the proof of the predecessor density estimate in [**AL1**] (see section II.6 for a discussion) apparently is motivated by the stochastic investigations in [**LaW**].

4. Related functions and generalizations

It is quite natural not to stay with the $3n+1$ function but to take into account other functions of "that type". A natural counterpart to the $3n + 1$ function is the following $3n - 1$ *function*

$$T' : \mathbb{N} \to \mathbb{N}, \qquad T'(n) = \begin{cases} n/2 & \text{if } n \text{ is even,} \\ (3n + 1)/2 & \text{if } n \text{ is odd,} \end{cases}$$

which is just the $3n + 1$ function T defined in (1.2), "on the negative integers": $T'(n) = -T(-n)$ for any $n \in \mathbb{N}$. Playing with $3n - 1$ trajectories leads to the

$3n - 1$ CONJECTURE. *For any starting number in* $\mathbb{N}$*, the* T'*-trajectory eventually ends in one of the following figures:*

   (1) *in the fix-point* (1),
   (2) *in the cycle* (5, 7, 10),
   (3) *or in the cycle* (17, 25, 37, 55, 82, 41, 61, 91, 136, 68, 34).

This conjecture, being as unsolved as the $3n+1$ conjecture itself, is considered by some authors including K. R. Matthews [**Mat2**] and B. G. Seifert [**Sei**].

A direct generalization of the $3n + 1$ function is

$$T_q : \mathbb{N} \to \mathbb{N}, \qquad T_q(n) = \begin{cases} n/2 & \text{if } n \text{ is even,} \\ (qn + 1)/2 & \text{if } n \text{ is odd,} \end{cases}$$

for odd integers $q > 1$, which may be called $qn + 1$ *function*. Z. Franco and C. Pomerance [**FP**] (1995) showed that, if q is what they call a Wieferich number, then there are natural numbers x with $T_q^k(x) \neq 1$ for any $k \in \mathbb{N}_0$. Here an odd integer q is called a *Wieferich number*, if $2^{\ell(q)} \equiv 1 \mod q^2$ where $\ell(q)$ is the order of 2 in the multiplicative group $(\mathbb{Z}/q\mathbb{Z})^*$. Moreover, they prove that the Wieferich numbers have relative asymptotic density 1 in the odd numbers. In 1978, R. E. Crandall [**Cra**] conjectured that *any* odd integer $q \geqslant 5$ has the property that is an $x_q \in \mathbb{N}$ such that $T_q^k(x_q) \neq 1$ for any $k \in \mathbb{N}_0$; we give some further remarks to that in section III.5.

Lagarias reports in [**Lag1**] that Collatz wrote in his notebook dated July 1, 1932, the integer function

$$(4.1) \qquad g(n) := \begin{cases} \frac{2}{3}n & \text{if } n \equiv 0 \mod 3, \\ \frac{4}{3}n - \frac{1}{3} & \text{if } n \equiv 1 \mod 3, \\ \frac{4}{3}n + \frac{1}{3} & \text{if } n \equiv 2 \mod 3, \end{cases}$$

which generates the cycles (1), $(2, 3)$, and $(4, 5, 7, 9, 6)$. It is not known whether the trajectory $(8, 11, 15, 10, 13, 17, 23, 31, 41, 55, 73, 97, 129, 86, 115, \ldots)$ is divergent, or whether it is eventually cyclic. This function is also discussed by D. Gale [**Gal**] (1991).

4.1. DEFINITION. Let $p \geqslant 2$ be an integer, and let $a_0, \ldots, a_{p-1}, b_0, \ldots, b_{p-1}$ be rational numbers subject to the $2p$ conditions

$$a_j p + b_j, a_j j + b_j \in \mathbb{Z} \qquad \text{for} \quad j = 0, \ldots, p - 1.$$

Then a *periodically linear function* is given by

$$U : \mathbb{Z} \to \mathbb{Z}, \qquad U(n) := a_j n + b_j \quad \text{if } n \equiv j \mod p.$$

These objects have been described by J. H. Conway [**Con**] (1972), in less formal terms by Lagarias [**Lag1**] (1985), and under the name *Collatz-like functions* by P. Michel [**Mic**] (1993).*

Conway proved in [**Con**] that there exists a periodically linear function $g : \mathbb{N} \to \mathbb{N}$ with the property that the behaviour of the g-iterates encodes the halting problem for Turing machines (see [**Lag1**], Theorem P, for a more precise statement of Conway's result). This implies that, in general, the limiting behaviour of the g-trajectories may be algorithmically undecidable. In fact, Conway describes a procedure how to construct a periodically linear function g with $b_0 = \cdots = b_{p-1} = 0$ whose dynamical behaviour encodes a previously fixed partial recursive function. On the other hand, if we fix a periodically linear function, say the $3n + 1$ function defined in (1.2), then it is not clear whether or not it is algorithmically decidable whether or not every T-iteration eventually runs into the cycle $(1, 2)$.

---

*I am grateful to M. Kudlek for drawing my attention to that paper.

S. Burckel [**Bur**] also presents Conway's result, gives more undecidability properties, and applies these ideas to problems about functional equations, see also section 13.

There is yet another link between periodically linear functions and Turing machines through the "busy beaver competition," as pointed out by P. Michel [**Mic**]. A "busy beaver" is a Turing machine which halts taking much time or leaving many marks, when starting from a blank tape. Michel analysed the halting problem of record holders in the five-state busy beaver competition and showed that this problem is connected to the dynamical behaviour of certain periodically linear functions.

There is a series of papers of K. R. Matthews [**Mat2**] and people around him, investigating the following mappings (to be notationally closer to this series of papers, we refrain for short from the reservation of T to denote the $3n + 1$ function):

4.2. DEFINITION. Let $d \geqslant 2$ be an integer, and let $m_0, \ldots, m_{d-1}$ be non-zero integers. Also for $i = 0, \ldots, d - 1$, let $r_i \in \mathbb{Z}$ satisfy $r_i \equiv im_i \mod d$. Then the formula

$$T(x) := \frac{m_i x - r_i}{d} \quad \text{if } x \equiv i \mod d .$$

defines a mapping $T : \mathbb{Z} \to \mathbb{Z}$ called a *generalized Collatz mapping*.

This definition is easily seen to be equivalent to definition 4.1, whence *periodically linear function* and *generalized Collatz mapping* are just two names for the same thing.

The paper [**MW1**] (1984) of K. R. Matthews and A. M. Watts deals with the *relatively prime case*, i.e. they suppose that $\gcd(m_i, d) = 1$ for $i = 0, \ldots, d - 1$. They start proving some interesting results about divergent T-trajectories, and they show that the extension $T : \mathbb{Z}_d \to \mathbb{Z}_d$ of a generalized Collatz mapping to the ring of d-adic integers is measure-preserving w.r.t. the Haar measure. Matthews and Watts extended their work in the direction of Markov matrices and ergodic theory in [**MW2**] (1985), to achieve results applying to a slightly more general case than the relatively prime case. In [**Lei**] (1986), G. M. Leigh abandons all restrictions on the moduli $m_0, \ldots, m_{d-1}$, except that they have to be non-zero. Like Matthews and Watts, also Leigh is mainly interested in divergent T-trajectories. He discovers a Markov chain (with, in some cases, infinite state space) closely related to the process of iterating T, and he shows that most of the results of [**MW2**] follow from his more general approach. More informations concerning ergodic properties of divergent T-trajectories have been given by R. N. Buttsworth and K. R. Matthews in [**BuM**] (1990). Ergodic properties of periodically linear functions are also studied by G. Venturini in a series of papers: [**Ven1**] (1982), [**Ven3**] (1992), [**Ven4**] (1997).

Matthews [**Mat1**] (1992) continued this work extending a generalized Collatz mapping to a continuous mapping on the polyadic numbers $\widehat{\mathbb{Z}}$, and dealing with the limiting frequencies in congruence classes. These results also apply mainly to divergent T-trajectories.

Analogues of generalized Collatz mappings on the ring $F_q[x]$ of polynomials over a finite field F_q have been investigated by Matthews and Leigh in [**ML**] (1987).

Intermediate between the $3n+1$ function T of (1.2) (from now on, the letter T will be used exclusively to denote that function) and periodically linear functions is *Hasse's generalization*:

4.3. DEFINITION. Let $d, m \in \mathbb{N}$ satisfy $d \geqslant 2$ and $\gcd(d, m) = 1$, assume that $A_d := \{0, r_1, \ldots, r_{d-1}\}$ is a complete system on incongruent residues to modulus d, and denote by $\varphi : \mathbb{Z} \to A_d$ the natural projection. Then a *Hasse function* is given by

$$H : \mathbb{Z} \to \mathbb{Z}, \qquad H(x) := \begin{cases} \dfrac{x}{d} & \text{if } x \equiv 0 \mod d, \\[2mm] \dfrac{mx - \varphi(mx)}{d} & \text{otherwise.} \end{cases}$$

Such Hasse functions have been considered by H. Möller [**Möl**] (1978), E. Heppner [**Hep**] (1978), and J.-P. Allouche [**All**] (1979). Some of their results are briefly described in section 6.

The functions considered in [**Wig**] (1988) and [**Ash**] (1992) are similar to these Hasse functions. These papers give some empirical observations concerning the length of cycles.

Another generalization of the $3n+1$ problem has been proposed by F. Mignosi [**Mig**] (1995), see also [**Bro**] (1995):

4.4. DEFINITION. Let $\alpha, r \in \mathbb{R}$ with $1 < \alpha < 2$, and put

$$T_{\alpha,r} : \mathbb{N} \to \mathbb{N}, \qquad T_{\alpha,r}(n) := \begin{cases} \lfloor \alpha n + r \rfloor & \text{if } n \text{ is odd,} \\ n/2 & \text{if } n \text{ is even.} \end{cases}$$

The set of purely periodic points of $T_{\alpha,r}$ is

$$L_{\alpha,r} := \left\{ k \in \mathbb{N} : \text{ there is a } k \in \mathbb{N} \text{ with } T_{\alpha,r}^k(n) = n \right\}.$$

Taking $\alpha = \frac{3}{2}$ and $r = \frac{1}{2}$, we see that the $3n+1$ function T is included in this class. *Mignosi's generalization* of the $3n+1$ conjecture reads (see [**Bro**]):

4.5. CONJECTURE. *The set $L_{\alpha,r}$ of cycles of $T_{\alpha,r}$ is finite, and for every integer $n \in \mathbb{Z}$ there is a $k \in \mathbb{N}$ with $T_{\alpha,r}^k(n) \in L_{\alpha,r}$.*

S. Brocco reports in [**Bro**] (1995) that this conjecture is proved for $r = 1$ and $\alpha = 2^{1/n}$ where $n \geqslant 2$, and also for $\alpha = 2^{2/3}$. On the other hand, Brocco shows that conjecture 4.5 is false if the following two conditions are satisfied:

(1) The interval $](r-1)/(\alpha-1)[$ contains an odd integer.

(2) α is a *Salem number* or a *Pisot-Vijayaraghavan number*.

Here a *Salem number* is a real algebraic integer > 1 all of whose conjugates are $\leqslant 1$ in modulus, one at least having modulus 1. A *Pisot-Vijayaraghavan number* is a real algebraic integer > 1 all of whose conjugates are < 1 in modulus.

Let us close this section with a few remarks about the entertaining combinatorics given by S. Anderson [**And**] (1987). He states a relation between the function $\widehat{f}$ of (2.1) and the sequence of Fibonacci numbers. Moreover, he claims that the function

$$g_a(x) := \begin{cases} x + a & \text{if } x \text{ is odd,} \\ x/2 & \text{if } x \text{ is even,} \end{cases}$$

for a fixed odd prime a, is related to a Fibonacci- (or Lucas-) type sequence. Finally, S. Anderson proposes not to study the $3n + 1$ function but instead the function

$$f_3(x) := \begin{cases} x/3 & \text{if } x \text{ is divisible by 3,} \\ x/2 & \text{if } x \text{ is divisible by 2 but not by 3,} \\ 3x + 1 & \text{otherwise,} \end{cases}$$

and generalizations thereof to other odd primes than 3.

5. Some formulae describing the iteration

In order to get a feeling for the mathematical content of $3n + 1$ iterations, let's consider some of the formulae associated to that iteration appearing in the literature. Of course, any author needs formulae; in this section, we just mention some striking examples which are useful but do not appear elsewhere in this chapter.

As a first instance, R. Terras [**Ter1**] (1976) uses the *parity function*, $X(n) := 1$ if n is odd, and $X(n) := 0$ if n is even, and rewrites (1.2) to

$$T(n) = \frac{1}{2} \left(3^{X(n)} n + X(n) \right).$$

Moreover, he defines $X_k(n) := X(T^k(n))$ for each $k \in \mathbb{N}_0$, $S_k(n) := X_0(n) + \cdots + X_{k-1}(n)$ (with $S_0(n) := 0$), and $\lambda_k(n) := 2^{-k} 3^{S_k(n)}$. From this he derives the formula

$$(5.1) \quad T^k(n) = \lambda_k(n)\, n + \varrho_k(n), \qquad \text{where} \quad \varrho_k(n) = \frac{\lambda_k(n)}{2} \sum_{j=0}^{k-1} \frac{X_j(n)}{\lambda_{j+1}(n)}.$$

To fix this idea, let us call $\lambda_k(n)$ the *k-th forward coefficient* of n, and $\varrho_k(n)$ the *k-th forward remainder* of n; note that the forward remainder is always non-negative, and it vanishes if and only if 2^k divides n. Backward analogues of these quantities, defined in definition 2.12, will play an important rôle in our approach.

A different way to write down multiple-step iterationshas been given by C. Böhm and G. Sontacchi [**BöS**] (1978). Put for the moment

$$f(x) := \frac{x}{2} \quad \text{and} \quad g(x) := \frac{3x + 1}{2} \quad \text{for } x \in \mathbb{Q},$$

and let $n_0, \ldots, n_m \in \mathbb{N}_0$ be given. Then

$$(5.2) \qquad f^{n_m} g f^{n_{m-1}} g \ldots f^{n_1} g f^{n_0}(x) = \frac{1}{2^n} \left(3^m x + \sum_{k=0}^{m-1} 3^{m-k-1} 2^{v_k} \right),$$

where $v_i = i + \sum_{j=0}^{i} n_i$ and $n = v_m$. Hence, the $3n + 1$ conjecture is equivalent to the following:

5.1. CONJECTURE. *Every natural number x is expressible in the form*

$$x = \frac{1}{3^m} \left(2^{v_m} - \sum_{k=0}^{m-1} 3^{m-k-1} 2^{v_k} \right),$$

where m is a positive integer, and $0 \leqslant v_0 < v_1 < \cdots < v_m$ are appropriate integers.

A nice non-branched formulafor the Collatz function f of (1.1) has been given by G. Meinardus [**Mei**] (1987):

$$f(n) = \frac{1}{4} \left((7n + 2) - (-1)^n (5n + 2) \right).$$

Using discrete Fourier transformation, he also gives a formula describing the m-step iteration of the Collatz function f: There are uniquely determined complex numbers $\alpha_{m,\mu}$, $\beta_{m,\mu}$, for $\mu = 0, \ldots, 2^m - 1$ such that

$$f^m(n) = \frac{1}{4^m} \sum_{\mu=0}^{2^m - 1} (\alpha_{m,\mu} n + \beta_{m,\mu}) \zeta_m^{\mu n},$$

where $\zeta_m = \exp(2\pi i / 2^m)$ denotes this 2^m-th root of unity.

The non-branched 1-step formula also has been rewritten for the $3n + 1$ function by L. Berg and G. Meinardus [**BeM**] (1994):

$$T(n) = \frac{1}{4} \left((4n + 1) - (-1)^n (2n + 1) \right).$$

A different type of a formula is given in the following

5.2. THEOREM. *(J. R. Kuttler [**Kut**], 1994) Let $k, n \in \mathbb{N}$, and $p \in \{1, \ldots, k\}$. Then there are exactly $\binom{k-1}{p-1}$ odd numbers $r \in \{1, 3, \ldots, 2^k - 1\}$ such that*

$$T^k(2^k n + r) = 3^p(n + \sigma)$$

for some rational σ with $0 < \sigma < 1$.

6. Numbers with finite stopping time

According to R. Terras [**Ter1**] (1976), the *stopping time* of $n \in \mathbb{N}$ is defined by

$$\sigma(n) := \begin{cases} \min\{k \in \mathbb{N} : T^k(n) < n\} & \text{if this set is non-empty,} \\ \infty & \text{if } T^k(n) \geqslant n \text{ for each } k \in \mathbb{N}. \end{cases}$$

(Actually, Terras writes χ instead of σ. Here we prefer σ which is more intuitve and which is used in [**Lag1**] and [**LeV**].) He uses his formula (5.1) to define the τ-*stopping time* (later called *coefficient stopping time* by J. C. Lagarias [**Lag1**]):

$$\tau(n) := \begin{cases} \min\{k \in \mathbb{N} : \lambda_k(n) < 1\} & \text{if this set is non-empty,} \\ \infty & \text{if } \lambda_k(n) \geqslant 1 \text{ for each } k \in \mathbb{N}. \end{cases}$$

As $\varrho_k(n) \geqslant 0$ in (5.1), we infer that $T^k(n) < n$ implies $\lambda_k(n) < 1$ which gives $\sigma(n) \geqslant \tau(n)$. Concerning the estimate in the other direction, Terras formulates the

6.1. COEFFICIENT STOPPING TIME CONJECTURE. *For all* $n \geqslant 2$, $\sigma(n) = \tau(n)$.

The main result in [**Ter1**] and [**Ter2**] is that the limits

$$F(k) := \lim_{m \to \infty} \frac{1}{m} |\{n \leqslant m : \sigma(n) \leqslant k\}|$$

exist for each $k \in \mathbb{N}$, and fulfill $\lim_{k \to \infty} F(k) = 1$. In 1977, C. J. Everett published a shorter proof of this result relying on Terras' encoding (see II.2.1) and using estimates from probability theory [**Eve**]. D. Boyd [**Boyd**] (1985) observed that there is a relation between $3n + 1$ iterations and certain *Pisot sequences* via a basic lemma of Terras.

The rate of convergence of the $F(k)$ has been estimated by Lagarias [**Lag1**]:

6.2. THEOREM. *For all integers* $k \geqslant 1$,

$$1 - F(k) \leqslant 2^{-\eta k},$$

where $\eta = 1 - H(\theta) \approx 0.05004$, *with* $H(x) = -x \log x - (1 - x) \log(1 - x)$ *and* $\theta = (\log_2 3)^{-1}$.

It is possible to use this result to obtain an estimate concerning numbers which do not have finite stopping time.

6.3. THEOREM. [**Lag1**] *There is a positive constant* c_1 *such that*

$$|\{n \in \mathbb{N} : n \leqslant x, \sigma(n) = \infty\}| \leqslant c_1 x^{1-\eta}.$$

Note that this theorem implies that, if a divergent T-trajectory happens to exist, it cannot diverge too slowly. Generalizations and sharpenings of these results to certain Hasse functions (see definition 4.3) have been given by several authors.

6.4. THEOREM. *Let H be a Hasse function with $m < d^{d(d-1)}$. Then:*

(a) *H. Möller* [**Möl**] *(1978):*

$$\lim_{x \to \infty} \frac{1}{x} \big| \{ n \leqslant x : H^k(n) < n \text{ for some } k \in \mathbb{N} \} \big| = 1 .$$

(b) *E. Heppner* [**Hep**] *(1978): there are real numbers $\delta_1, \delta_2 > 0$ such that, for $N = \lfloor \log_d x \rfloor$, we have*

$$\big| \{ n \leqslant x : H^N(n) \geqslant n \, x^{-\delta_1} \} \big| = O(x^{1-\delta_2}) .$$

Note that the $3n+1$ function T is included in the class of functions considered in this theorem. Indeed, T is the Hasse function with $m = 3$, $d = 2$, and $A_d = \{0, 1\}$, satisfying clearly $m = 3 < 2^2 = d^{d(d-1)}$.

A keyword to the method used by Möller is "F-Normalreihen." Heppner's article is, in a sense, an answer to Möller's announcement; Heppner uses probability methods to achieve his result, which is, in fact, considerable stronger than Möller's. A similar result as Heppner's has been obtained, with different methods, by J.-P. Allouche [**All**] (1979). As far as the $3n + 1$ problem itself is concerned, the best result (up to now) in this direction is the following theorem of I. Korec [**Kor2**] (1994), which again has been proved using arguments from probability analysis.

6.5. THEOREM. *For every real $c > \log_4 3$ the set*

$$M_c = \{ y \in \mathbb{N} : \text{ there is an } n \in \mathbb{N} \text{ such that } T^n(y) < y^c \}$$

has asymptotic density one, i.e. the limit $\displaystyle \lim_{x \to \infty} \frac{|\{ y \in M_c : y \leqslant x \}|}{x}$ *exists and is equal to one.*

A lower bound for such a constant c also is constructed in [**All**], but there the lower bound is $\frac{3}{2} - \log_3 2 \approx 0.8691 > \log_4 3 \approx 0.7925$.

Another result, different on a first look, has been obtained by J. M. Dolan, A. F. Gilman, and S. Manickam [**DGM**] (1987).

6.6. THEOREM. *Let $k \in \mathbb{N}$. Then the set*

$$\{ m \in \mathbb{N} : \text{ there is an } n \in \mathbb{N} \text{ such that } T^{n+p}(m) < m \text{ for } p \in \{0, \dots, k-1\} \}$$

has asymptotic density one.

The following theorem of G. Venturini [**Ven2**] (1989) is noted here for completeness; it is, in fact, an immediate corollary of Heppner's result, but it easily implies theorem 6.6.

6.7. THEOREM. *For every real number $\rho > 0$, the set*

$$\{ m \in \mathbb{N} : \text{ there is an } n \in \mathbb{N} \text{ such that } T^n(m) < \rho m \}$$

has asymptotic density one.

7. Asymptotics of predecessor sets

A much stronger question than that rised in the preceding section is the following one: What can be said about predecessor sets, i.e. about the sets

$$\mathcal{P}_T(a) := \{ n \in \mathbb{N} : \text{ there is a } k \geqslant 0 \text{ such that } T^k(n) = a \},$$

where $a \in \mathbb{N}$ is given? Of special interest are estimates concerning the asymptotic behaviour, as $x \to \infty$, of the counting function

$$Z_a(x) := Z_{\mathcal{P}_T(a)}(x) := \left| \{ n \in \mathcal{P}_T(a) : n \leqslant x \} \right|.$$

A first result in this dierection has been given by R. E. Crandall [**Cra**] (1978; th. 6.1):

7.1. THEOREM.   *There is a positive constant c such that $Z_1(x) > x^c$ for sufficiently large $x \in \mathbb{N}$.*

J. W. Sander computed this constant to $c \approx 0.057$ and improved it to $c = 1/4$ [**San**] (1990; the results already had been presented during a conference in 1987). This "tree-search method" has been improved by D. Applegate and J. C. Lagarias [**AL1**] (1995), arriving at a computer-assisted proff for $c = 0.654$. We discuss the tree-search method in detail in section II.6.

Using a different approach, I. Krasikov [**Kra**] (1989) succeeded in proving $Z_a(x) > x^\beta$ with $\beta = 3/7$, for sufficiently large x and certain numbers $a \in \mathbb{N}$ (including $a = 1$). This method can be refined to yield $\beta = 0.48$ [**Wir1**] (1993). The technique of "Krasikov inequalities" also has been improved by D. Applegate and J. C. Lagarias [**AL2**] (1995), Based on solving certain nonlinear programming problems, they obtain a computer-assisted proof of the following estimate.

7.1. THEOREM.   *For each $a \not\equiv 0 \mod 3$, there is a positive constant c_a such that*

$$Z_a(x) \geqslant c_a\, x^{0.81} \qquad \text{for all} \quad x \geqslant a.$$

8. Consecutive numbers with the same height

According to L. E. Garner [**Gar**] (1984), the *height* of a natural number n is defined to be the least non-negative integer k such that $T^k(n) = 1$ (if such an integer exists). It is necessary to distinguish two height functions corresponding to the Collatz function f defined in (1.1) and the $3n + 1$ function T defined in (1.2), respectively. We put

$$h_C(n) := \min \{ k \in \mathbb{N}_0 : f^k(n) = 1 \}$$

and

$$h_T(n) := \min \{ k \in \mathbb{N}_0 : T^k(n) = 1 \},$$

where the min of an empty set is understood to be ∞.

Several people observed that there occur long runs of integers having the same height (this occurs with both height functions h_C and h_T).

8.1. THEOREM. *(L. E. Garner [Gar], 1984) There are infinitely many pairs of consecutive positive integers $(n, n+1)$ whose T-trajectories coincide at a certain step, i.e. there is a non-negative integer k with the property $T^k(n) = T^k(n+1)$.*

In fact, Garner proves much more, giving explicit constructions and patterns in the corresponding parity vectors (cf. definition II.2.1). Theorems about consecutive numbers of the same height are also given by P. Filipponi [Fil] (1991).

Longer runs of numbers with the same height (now w.r.t. the Collatz function) have been observed by P. Penning [Pen] (1989). J. B. Wu [Wu1] (1992) asks for the number $K(N)$ of integers in the interval $[1, 2^N)$ which belong to an n-tuple $(n \geq 2)$ of consecutive numbers with the same height. He calculates $K(N)$ for $N = 1, \ldots, 24$ and claims that the longest such n-tuple in the interval $[1, 2^{30})$ is the 176-tuple starting with 722 067 240.

A better result in this direction is

8.2. THEOREM. *(G.-G. Gao [Gao], 1993) There are infinitely many 35 654-tuples of consecutive positive integers such that all of the integers in one tuple have the same height h_C.*

9. Cycles

There is a certain amount of beautiful results in the $3n+1$ literature concerning cycles. J. L. Davison [Dav] (1976) called an iteration of the $3n+1$ function a *circuit*, if it looks like

$$\left(n \xrightarrow{k} m \xrightarrow{\ell} n^* \right) := \left(n, T(n), \ldots, T^k(n) = m, T(m), \ldots, T^\ell(m) = n^* \right)$$

with $n < T(n) < \cdots < T^k(n)$ and $m > T(m) > \cdots > T^\ell(m)$, and he derives for such a circuit the formulae $2^k(m+1) = 3^k(n+1)$ and $n^* = m/2^\ell$. Then he investigates circuits which are cycles ($n = n^*$), and characterizes these objects by relating them to solutions of a certain exponential Diophantine equation.

9.1. THEOREM. *(J. L. Davison) There is a one-one-correspondence between $3n+1$ circuits which are cycles, and triples $(k, \ell, h) \in \mathbb{N}^3$ solving*

$$(2^{k+\ell} - 3^k)h = 2^\ell - 1 .$$

R. P. Steiner [Ste1] (1978) proved that this exponential Diophantine equation has no other solution than $(k, \ell, h) = (1, 1, 1)$, using continued fractions and a lemma of A. Baker on linear forms in logarithms. This implies that the known $3n+1$ cycle $(1, 2)$ is the only one that is also a circuit. A slightly simpler proof of this has been given by O. Rozier [Ro1] (1990), using a result of M. Waldschmidt.

In [Ste2] and [Ste3], Steiner transferred his method to the more general $qn+1$ problem, finding that for $q = 5$ there is only one "non-trivial" circuit-cycle $13 \xrightarrow{3} 208 \xrightarrow{4} 13$ (in Davison's notation), whereas $q = 7$ admits no "non-trivial" circuit-cycles. (Steiner defines the $qn+1$ function such that 1 is a fixed point which is a "trivial" cycle, so he didn't mention the circuit-cycles $1 \xrightarrow{2} 8 \xrightarrow{3} 1$ for $q = 5$ and $1 \xrightarrow{1} 4 \xrightarrow{2} 1$ for $q = 7$.)

Continued fractions also have been used to obtain informations about $3n + 1$ cycles with a more complicated structure than simple circuit-cycles, by several authors: R. E. Crandall [Cra] (1978), B. Schuppar [Sch] (1981), and S. Eliahou [Eli] (1994). The latter author gave an interesting condition on the length of a cycle not falling under a certain minimum.

9.2. THEOREM. *(S. Eliahou) Let Ω be a non-trivial cycle of T. Provided* $\min \Omega > 2^{40}$, *we have*

$$|\Omega| = 301\,994\,a + 17\,087\,915\,b + 85\,137\,581\,c\,,$$

where a, b, c are non-negative integers, $b > 0$, $ac = 0$.

Applying results from the theory of linear forms in logarithms, finiteness results on cycles of certain types have been obtained by O. Rozier [Ro2] (1991), and E. Belaga and M. Mignotte [BMi] (1996).

C. Böhm and G. Sontacchi [BöS] (1978) derive from their m-step formula (5.2) the following cycle condition:

9.3. THEOREM. *Let $x \in \mathbb{Z}$. Then there is an $n \in \mathbb{N}$ with $T^n(x) = x$, if and only if there are integers $0 \leqslant v_0 < v_1 < \cdots < v_m = n$ such that*

$$x = \frac{1}{2^n - 3^m} \sum_{k=0}^{m-1} 3^{m-k-1} 2^{v_k}\,.$$

B. G. Seifert [Sei] (1988) used similar formulae to express the cycle condition in terms of a group action of a finite cyclic group. He points out the following consequence of his results: The $3n + 1$ conjecture and the $3n - 1$ conjecture imply, respectively, that the equations

$$(9.1) \qquad\qquad 2^\ell - 3^r = 1\,, \qquad 2^\ell - 3^r = -1$$

have only the positive integral solutions

$$(9.2) \qquad (\ell, r) = (2, 1)\,, \qquad (\ell, r) = (1, 1) \text{ or } (3, 2)\,, \qquad \text{respectively.}$$

But this statement is void, as it is not difficult to see that the equations (9.1) only have the solutions (9.2). Indeed*, a computation modulo 3^r shows that the first equation of (9.1) implies $2^\ell \equiv 1 \mod 3^r$. By well-known facts concerning the group of prime residues to modulus 3^r (see, for instance, [Has]), we infer that this means $\ell \equiv 0 \mod 2 \cdot 3^{r-1}$. Because $\ell = 0$ does not give a solution, we have the implication

$$2^\ell - 3^r = 1 \quad \Longrightarrow \quad \ell \geqslant 2 \cdot 3^{r-1}\,.$$

---

*For this observation, I am grateful to V. Krafft [Krf].

This implies the middle ">" in the chain

$$3^r - 1 = 2^\ell \geqslant 2^{2 \cdot 3^{r-1}} > 3^r - 1 \quad \text{for } r > 1,$$

where the last inequality follows by a simple computation using $r > 1$. Hence, any solution of $2^\ell - 3^r = 1$ must fulfill $r = 1$, which implies that $(\ell, r) = (2, 1)$ is the only solution of this equation.

Similarly, we obtain the implications

$$2^\ell - 3^r = -1 \quad \Longrightarrow \quad \ell \equiv 3^{r-1} \mod 2 \cdot 3^{r-1} \quad \Longrightarrow \quad \ell \geqslant 3^{r-1},$$

which, in turn, imply the chain

$$3^r + 1 = 2^\ell \geqslant 2^{3^{r-1}} > 3^r + 1 \quad \text{for } r > 2.$$

As before, the last inequality is a simple computation using $r > 2$. This proves that the second equation $2^\ell - 3^r = -1$ admits in natural numbers only the solutions $(\ell, r) = (1, 1)$ and $(\ell, r) = (3, 2)$.

J. C. Lagarias [**Lag2**] (1991) considered the natural extension of the $3n + 1$ function T to the ring $\mathbb{Q}[(2)]$ of rational numbers with odd denominator. (Note that $\mathbb{Q}[(2)]$ is a subring of the ring $\mathbb{Z}_2$ of 2-adic integers, cf. section 10.) This extension admits a large set of cycles, which is studied in [**Lag2**]. These *rational cycles* of the $3n + 1$ function can be represented as certain cycles of $3n + k$ functions, with $k \geqslant 1$ and $k \equiv \pm 1 \mod 6$. Rational cycles of T are also studied by B. P. Chisala [**Chi**] (1994).

10. Binary sequences and 2-adic analysis

The binary representation of a natural number n ends on 1, if n is odd, and it ends on 0, if n is even. This simple fact is the motivation to investigate the behaviour of the $3n + 1$ function T (or, almost equivalently, the Collatz function) on binary sequences. In this setting, natural numbers are represented either by finite binary sequences, or by infinite sequences containing only finitely many 1's.

Let us first briefly consider papers on the $3n + 1$ problem dealing with finite binary sequences. J. Błazewicz and A. Pettorossi [**BłP**] (1983) describe the Collatz function (and also the $3n+1$ function) in terms of the theory of *Rewriting Systems* and prove some nice theorems concerning the past and future of binary sequences containing runs of 0's or runs of 1's. Related to this are also an article by J. O. Shallit [**Sha**] (1991), and a result obtained by D. W. Wilson [**Wls**] (1991). T. Cloney, E. Goles, and G. Y. Vichniac [**CGV**] (1987) represent the $3n + 1$ function in base 2 as a quasi cellular automaton. They do not so much prove theorems but present nice computer graphics showing patterns and chaotic behaviour of $3n + 1$ iterations operating on finite binary sequences.

From a mathematical point of view, the extension of the $3n+1$ function to the ring $\mathbb{Z}_2$ of 2-adic integers, which can be represented by *infinite* binary sequences, is more interesting. A 2-adic number $a \in \mathbb{Z}_2$ corresponding to a binary sequence $(a_j)_{j \in \mathbb{N}_0}$, $a_j \in \{0, 1\}$, is usually written as a sum

$$a = a_0 + a_1 2 + a_2 2^2 + \cdots = \sum_{j=0}^{\infty} a_j 2^j \, .$$

The notation of congruences modulo 2 is also used on $\mathbb{Z}_2$, writing $a \equiv a_0$ mod 2. $\mathbb{Z}_2$ is a complete metric space, with the 2-adic metricinduced by the 2-adic valuation; it is a topological group admitting a unique normalized Haar measure; moreover, differentiability and analyticity are naturally defined for functions on $\mathbb{Z}_2$, see, for instance, the book of K. Mahler [**Mah**]. Then the extension of the $3n + 1$ function to $\mathbb{Z}_2$ is given by

(10.1) $T : \mathbb{Z}_2 \to \mathbb{Z}_2$, $T(a) := \begin{cases} a/2 & \text{if } a \equiv 0 \mod 2 \, , \\ (3a + 1)/2 & \text{if } a \equiv 1 \mod 2 \, . \end{cases}$

The following theorem combines results of K. R. Matthews and A. M. Watts [**MW1**] (1984) and H. Müller [**Mül1**] (1991).

10.1. THEOREM. *The above map* $T : \mathbb{Z}_2 \to \mathbb{Z}_2$ *is surjective, not injective, infinitely many times differentiable, not analytic, measure-preservingw.r.t. the Haar measure on* $\mathbb{Z}_2$, *and strongly mixing.*

Ergodic properties of T on the set of infinite binary sequences, without using the algebraic structure of $\mathbb{Z}_2$, have also been observed by G. Keller [**Kel**] (1984)*.

To obtain more information about the $3n+1$ problem, let us follow H. Müller's approach [**Mül2**] (1994) and consider the following infinite matrices, and their *column functions*.

10.2. DEFINITION. Let $T : \mathbb{Z}_2 \to \mathbb{Z}_2$ be as in (10.1), and let $a \in \mathbb{Z}_2$. Then the coefficients $a_{ij} \in \{0, 1\}$ of the (infinite binary) matrix $M(a) := (a_{ij})_{i,j \in \mathbb{N}_0}$ are defined by

$$T^i(a) = \sum_{j=0}^{\infty} a_{ij} \, 2^j \, ,$$

where T^i denotes the i-fold iterate of T, with $T^0 = \text{id}$. For each $j \in \mathbb{N}_0$, the j-*th column function* is

$$Q_j : \mathbb{Z}_2 \to \mathbb{Z}_2 \, , \qquad Q_j(a) := \sum_{i=0}^{\infty} a_{ij} \, 2^i \, .$$

The essentials of the following theorem already occur in Terras' paper [**Ter1**] (1976), whereas the explicit statement is found in [**Lag1**] (1985). Actually, Lagarias does not use the term "column function," which has been fixed later by Müller; Lagarias writes Q_∞ instead of Q_0.

---

*I am grateful to K. P. Hadeler for sending me a copy of that letter.

10.3. THEOREM. *The column function* $Q_0 = Q_\infty : \mathbb{Z}_2 \to \mathbb{Z}_2$ *is continuous, bijective, and measure-preserving on* $\mathbb{Z}_2$.

Part of this theorem has been considerably sharpened and extended to the other column function by H. Müller in [**Mül1**] (1991) and [**Mül2**] (1994), to arrive at the following results.

10.4. THEOREM. *For each* $j \in \mathbb{N}_0$, *the column function* $Q_j : \mathbb{Z}_2 \to \mathbb{Z}_2$ *is continuous and nowhere differentiable.*

Bernstein [**Brs**] (1994) also proves non-differentiability of his function $\Phi = Q_\infty^{-1} = Q_0^{-1}$, using another method than that given by H. Müller.

10.5. THEOREM. *A linear combination*

$$f : \mathbb{Z}_2 \to \mathbb{Q}_2, \qquad f(a) = \sum_{j=0}^{N} A_j \, Q_j(a)$$

with coefficients $A_j \in \mathbb{Q}_2$ *is differentiable at a point* $a \in \mathbb{Z}_2$ *with* $T^{k_0}(a) = 0$ *for some* $k_0 \in \mathbb{N}_0$, *if, and only if,*

$$2A_0 + A_1 = 0, \qquad A_2 = A_3 = \cdots = A_N = 0.$$

H. Müller also noted in [**Mül2**] that the linear combination $Q_0 - 2Q_1$ is locally constant in $\mathbb{Z}_2$, showing thus that theorem 10.5 characterizes all linear combinations of column functions which are everywhere differentiable in $\mathbb{Z}_2$.

D. J. Bernstein [**Brs**] (1994) explicitly constructs the inverse of $Q_0 = Q_\infty$, which he denotes by $\Phi : \mathbb{Z}_2 \to \mathbb{Z}_2$. For the following non-iterative statements of the $3n + 1$ conjecture, observe that rational numbers with odd denominator are 2-adic integers, e.g. $\frac{1}{3}\mathbb{Z} \subset \mathbb{Z}_2$. The equivalence (a) $\Leftrightarrow$ (b) already has been observed by Lagarias [**Lag1**] (1985).

10.6. THEOREM. *The following assertions are equivalent:*

(a) *The* $3n + 1$ *conjecture is true.*
(b) $Q_\infty(\mathbb{N}) \subset \frac{1}{3}\mathbb{Z}$.
(c) $\mathbb{N} \subset \Phi(\frac{1}{3}\mathbb{Z})$.

More interesting facts about Bernstein's function Φ are given by Bernstein and Lagarias in [**BrsL**] (1996). Let $S : \mathbb{Z}_2 \to \mathbb{Z}_2$ denote the *2-adic shift map*:

$$S\left(\sum_{j=0}^{\infty} a_j 2^j\right) := \sum_{j=0}^{\infty} a_{j+1} 2^j.$$

Then it can be shown that Φ is uniquely determined by the properties

$$\Phi \circ S \circ \Phi^{-1} = T \qquad \text{and} \qquad \Phi(0) = 0.$$

Therefore, Φ is called $3x + 1$ conjugay map. Moreover, it is easy to check that Φ is *solenoidal*, i.e., Φ induces permutations Φ_n on $\mathbb{Z}/2^n\mathbb{Z}$. This connects the study of $3n + 1$ trajectories to the study of dynamical systems on the 2-adic integers. For example: A natural conjecture is (see [**Lag1**]):

PERIODICITY CONJECTURE. $\Phi(\mathbb{Q} \cap \mathbb{Z}_2) = \mathbb{Q} \cap \mathbb{Z}_2$.

Using arguments of [**Lag2**], it is proved in [**BrsL**] that this periodicity conjecture implies the divergent trajectory conjecture for $3n + 1$ iterations.

11. Reduction to residue classes and other sets

C. C. Cadogan [**Cad**] (1984) used an observation of Š. Znám to prove the following result; here f denotes the Collatz function (1.1).

11.1. THEOREM. *Let* $A_j := \left\{ x \in \mathbb{N} : x \equiv 2^{j-1} - 1 \mod 2^j \right\}$ *for integers* $j \geqslant$ 2. *Then* $j > 2$ *and* $x \in A_j$ *imply* $f^{2(j-2)}(x) \in A_2$.

In the note [**Pud**] (1986) of S. Puddu, it is shown that every natural number m has some iterate $f^k(m) \equiv 1 \mod 4$ (this also follows from theorem 11.1). If $m \equiv 3 \mod 4$, then the smallest such k must fulfill $f^k(m) \equiv 5 \mod 12$.*

In this vein, I. Korec and Š. Znám [**KoZ**] (1987) proved

11.2. THEOREM. *Let* p *be an odd prime such that 2 is a primitive root modulo* p^2, *and let* a, n *be positive integers with* $p \nmid a$. *Then, for all positive integers* x, *there are positive integers* b, i, j *such that* $b \equiv a \mod p^n$ *and* $f^i(x) = f^j(b)$.

Korec and Znám also claim that they have constructed a set $X \subset \mathbb{N}$ with vanishing asymptotic density such that, for all positive integers y, there are positive integers x, i, j with $x \in X$ and $f^i(y) = f^j(x)$.

12. Formal languages

There is an interesting paper of I. Korec [**Kor1**] (1992) connecting the $3n + 1$ problem to certain structural properties of certain finite algebras.

Korec considers algebras $\mathcal{A} = (A, \star, o)$ consisting of a finite set A, a binary operation $\star$ which is not assumed to be associative or commutative, and with a constant $o \in A$. To such an algebra he associates a family of functions (with values in A) called *generalized Pascal triangles*, abbreviated GPT. These functions had been used to study the structure of certain cellular automata. The GPT can happen to have some properties, e.g. they can be *nilpotent* or *simple semilinear of degree k* (these notions are defined in a very clear way in [**Kor1**]; the concepts are too complicated to repeat the formal definitions here), which are, in fact, structural properties of the underlying algebras.

In this setting, Korec constructs two such algebras $\mathcal{A}_1 = (A, *, o)$ and $\mathcal{A}_2 = (A, \oplus, o)$ with 7 and 8 elements, respectively. The main results in [**Kor1**] are

12.1. THEOREM. *The following assertions are equivalent:*

(a) *The* $3n + 1$ *conjecture holds.*
(b) *The algebra* $\mathcal{A}_1$ *is simple semilinear of degree 2.*
(c) *The algebra* $\mathcal{A}_2$ *is nilpotent.*

Korec also connects structural properties of his algebra $\mathcal{A}_1$ to two intermediate conjectures:

---

*This is cited after Lagarias [**Lag3**], as I did not manage to obtain a copy of Puddu's paper.

NO DIVERGENT TRAJECTORY CONJECTURE. *There is no divergent $3n + 1$ trajectory, i.e., there is no $y \in \mathbb{N}$ such that* $\lim\limits_{n \to \infty} T^n(y) = \infty$.

FINITE CYCLES CONJECTURE. *There are only finitely many $y \in \mathbb{N}$ such that $T^n(y) = y$.*

Of course, Korec's results can only be indicated here without giving the explicit constructions. But perhaps these few remarks could serve to encourage the interested reader to study Korec's paper [**Kor1**] (cf. also [**CGV**] for a less formal introduction into the subject).

13. Functional equations

There is a connection between the $3n + 1$ problem and holomorphic functions in the complex unit disc $\mathbb{E} = \{z \in \mathbb{C} : |z| < 1\}$ via functional equations for such functions. L. Berg and G. Meinardus [**BeM**] (1994), see also [**BeM2**] (1995), proved the following result:

13.1. THEOREM. *The following assertions are equivalent:*

(a) *The $3n + 1$ conjecture holds.*
(b) *For each $n \in \mathbb{N}$, the function*

$$g_n(w) := \sum_{m=0}^{\infty} T^m(n) \, w^m$$

is a rational function of the form $g_n(w) = \dfrac{q_n(w)}{1 - w^2}$, where q_n is a polynomial with integer coefficients.

(c) *Let ζ denote the third root of unity $\zeta = e^{2\pi i/3}$. The only solutions of the functional equation*

$$h(z^3) = h(z^6) + \frac{h(z^2) + \zeta h(\zeta z^2) + \zeta^2 h(\zeta^2 z^2)}{3z}$$

which are holomorphic in the unit disc $\mathbb{E}$ have the form $h(z) = h_0 + \dfrac{h_1 z}{1 - z}$ with arbitrary constants $h_0, h_1 \in \mathbb{C}$.

Basic ideas of the proof, and equivalent versions of these results for the Collatz function (1.1) instead of the $3n + 1$ function T of (1.2), already appear in [**Mei**] (1987).

Related to this theorem is a result of S. Burckel [**Bur**] (1994). To state it, consider the following set of entire functions:

$$\mathcal{S} := \left\{ f(z) = \sum_{n=0}^{\infty} s_n z^n : s_n \in \{0, 1\} \text{ for all } n \in \mathbb{N}_0 \right\}.$$

13.2. THEOREM. *(S. Burckel* [**Bur**]*, proposition 4.1) The $3n + 1$ conjecture is true if and only if the equation*

$$3z^3 R(z^3) - 3z^9 R(z^6) - R(z^2) - R(\omega z^2) - R(\omega^2 z^2) = 0$$

only admits the trivial solution in S*, where* $\omega = e^{2i\pi/3}$.

Note that Burckel's functional equation follows from the functional equation in (c) of theorem 13.1 by the substitution $h(z) = z^2 R(z)$.

14. A continuous extension to the real line

M. Chamberland [**Cha**] (1996) observed that the entire function defined by

$$f(x) = \frac{x}{2} \cos^2\left(\frac{\pi x}{2}\right) + \frac{3x+1}{2} \sin^2\left(\frac{\pi x}{2}\right) = x + \frac{1}{4} - \frac{2x+1}{4} \cos(\pi x)$$

interpolates the $3n + 1$ function T as defined in (1.2): $f(x) = T(x)$ for any $x \in \mathbb{N}$. This allows him to try to apply methods of one-dimensional discrete dynamical systems to $3n + 1$ iterations.

As a first step, Chamberland shows that the Schwarzian derivative of f,

$$Sf(x) = \frac{f'''(x)}{f'(x)} - \frac{3}{2}\left(\frac{f''(x)}{f'(x)}\right)^2 ,$$

fulfills $Sf(x) < 0$ for any real $x \geqslant 0$. This makes it possible to apply a couple of theorems concerning periodic points in one-dimensional discrete dynamics to f.

14.1. LEMMA.

(a) *Denote by* $\mu_0, \mu_1, \mu_2, \ldots$ *the fixed points of* f *on* $[0, \infty)$ *in increasing order. Then* $\mu_0 = 0$, $n - 1 \leqslant \mu_n \leqslant n$ *for any* $n \in \mathbb{N}$, *and*

$$\mu_n = n - \frac{1}{2} + O\left(\frac{1}{n}\right) \qquad \text{as } n \to \infty.$$

(b) *Denote by* $c_1, c_2, c_3, \ldots$ *the critical points of* f *on* $[0, \infty)$ *in increasing order. Then* $\mu_n < c_n < \mu_{n+1}$ *for any* $n \in \mathbb{N}$, *and*

$$c_n = n - \frac{2}{\pi^2 n}\left((-1)^n - \frac{1}{2}\right) + O\left(\frac{1}{n^2}\right) \qquad \text{as } n \to \infty.$$

14.2. THEOREM. *The intervals* $I_1 := [0, \mu_1]$ *and* $I_2 := [\mu_1, \mu_2]$ *are invariant under* f.

(a) I_1 *contains a single attracting fixed point* $\mu_0 = 0$, *and every point in* I_1 *is attracted to* 0 *except the repelling fixed point* μ_1.
(b) I_2 *contains exactly two attracting periodic orbits:*

$$A_1 = \{1, 2\}, \qquad A_2 = \{1.192531907\ldots, 2.138656335\ldots\}.$$

There is a partition $I_2 = \Omega(A_1) \cup \Omega(A_2) \cup \Gamma$ *in which* $\Omega(A_i)$ *is the basin of attraction for the set* A_i, $i = 1, 2$, *and* Γ *is a set of measure zero.*

The remaining interval $I_3 := [\mu_3, \infty)$, which is not invariant, is partitioned to $I_3 = E_f \cup R_f$, where $E_f := \{x \in I_3 : \limsup f^n(x) < \mu_3\}$ is the *escape set*, and $R_f := \{x \in I_3 : \liminf f^n(x) \geq \mu_3\}$ is the *residual set*. In this setting, the $3n+1$ conjecture reads:

$3n+1$ CONJECTURE. *(Chamberland)* $\quad \mathbb{N} \cap R_f = \varnothing$.

The residual set R_f is further partitioned to $R_f = S_f \cup U_f^0 \cup U_f^\infty$ where S_f is the *stable set*, U_f^0 consists of the unstable but bounded orbits, and U_f^∞ is the union of unbounded orbits. Now the divergent trajectory conjecture reads:

DIVERGENT TRAJECTORY CONJECTURE. *(Chamberland)* $\quad \mathbb{N} \cap U_f^\infty = \varnothing$.

Chamberland obtains some interesting results about the dynamical system on $\mathbb{R}$ generated by his entire function f:

14.3. THEOREM.

(a) *Any cycle of the $3n+1$ function T on $\mathbb{N}$ is an attractive cycle of f on $\mathbb{R}$.*
(b) *The set U_f^0 is uncountable.*
(c) *Given $\varepsilon > 0$, then*

$$(1, \infty) \subset \bigcup_{\mu_3 < x < \mu_3 + \varepsilon} \mathrm{Orb}(x) \; ;$$

here $\mathrm{Orb}(x) := \{f^n(x) : n \in \mathbb{N}_0\}$ *denotes the f-orbit of x.*
(d) *The set U_f^∞ contains a monotonically increasing divergent trajectory.*

This theorem shows that a thorough study of the dynamics of a continuous extension of the $3n+1$ function is not easier than the $3n+1$ problem itself.

CHAPTER II

ANALYSIS OF THE COLLATZ GRAPH

It is an idea of L. Collatz to represent a given function $g : \mathbb{N} \to \mathbb{N}$ by a directed graph with the natural numbers as vertices and edges $(n, g(n))$ [**Col1, Col2**]. We shall denote this graph by Γ_g; it has been called *Collatz graph* of g by J. C. Lagarias [**Lag1**]. The Collatz graph is especially well-suited for the study of the dynamical system generated by iterating the integer function, because there is a natural one-one-correspondence between the (directed) paths in the graph and the finite sequences arising from iterated applications of the given function.

In this chapter, we are mainly concerned with the Collatz graph of the $3n + 1$ function

$$T : \mathbb{N} \to \mathbb{N}, \qquad T(n) = \begin{cases} T_0(n) = n/2 & \text{if } n \text{ is even,} \\ T_1(n) = (3n + 1)/2 & \text{if } n \text{ is odd.} \end{cases}$$

Due to the form of the two branches of T, this Collatz graph shows a lot of structure and recurrent patterns which we shall investigate in some detail in this chapter.

The first section gives basic notions concerning directed graphs and describes how to represent functions on the natural numbers by directed graphs with the natural numbers as vertices. In this way, properties of the dynamical system induced by iterating the integer function are pictured in properties of the graph representing the function: for instance, a domain of attraction is represented by the vertices belonging to a weak component of the graph.

In the second section, we restrict our attention to the $3n + 1$ function and describe possible encodings of paths in the Collatz graph, or, equivalently, of finite portions of T-trajectories like

$$b \xrightarrow{T} T(b) \xrightarrow{T} T^2(b) \xrightarrow{T} \ldots \xrightarrow{T} a = T^n(b) \, .$$

As each step T must come from one of the two branches T_0 and T_1 of the $3n + 1$ function, the above path in the Collatz graph Γ_T gives rise to a T_0-T_1-sequence of length n. Representing each T_0 by 0 and each T_1 by 1, we obtain, according to R. Terras [**Ter1**], the *parity vector* of length n associated to b. Terras observed that the parity vector representing a path of length n with initial vertex b depends on b only through its residue class modulo 2^n; moreover, he established a one-one-correspondence between all possible 0-1-vectors of length n and the residue classes modulo 2^n. I do not rederive Terras' result but propose a different encoding of such paths by *encoding vectors* of non-negative integers, as follows. The number of entries in an encoding vector representing a path, say, from b to a of length n, is precisely one more than the number of edges in that path arising from T_1; each entry of the encoding vector to a path is given by the number of edges coming from T_0 between two consecutive occurrences of T_1. So the encoding vector $(s_0, \ldots, s_\ell)$, say, represents the following path of length $n = \ell + s_0 + \cdots + s_\ell$: the first s_0 edges come from T_0, the next one comes from T_1, then the next s_1 edges come from T_0, then again one from T_1, and so on, until the path ends with s_ℓ edges arising from T_1. If $s_0, \ldots, s_\ell$ run through the non-negative integers, we see that we can encode in this way all possible paths with precisely ℓ occurrences of T_1. An encoding vector will be called *admissible* w.r.t. a given vertex $a \in \mathbb{N}$, if it encodes a path in the Collatz graph with terminal vertex a. It will turn out later in this chapter that this encoding is especially well-suited for representing paths *terminating* at a given vertex. This encoding is, roughly speaking, in this respect dual to the encoding by parity vectors, as parity vectors are better suited to encode paths with given *initial* vertex.

The third section exhibits some basic properties of admissible integer vectors. Given a non-negative integer vector $s = (s_0, \ldots, s_\ell)$, we show that there is a residue class modulo 3^ℓ such that s is admissible w.r.t. a vertex $a \in \mathbb{N}$, if and only if a is a member of that residue class. We also discuss some consequences of this observation, e.g., there are certain infinite subgraphs of the Collatz graph of the $3n + 1$ function which recur infinitly many times.

Concerning the dynamics of $3n + 1$ iterations, informations about the size of a domain of attraction would be really interesting. More concretely, we are seeking information about the size of predecessor sets $\mathcal{P}_T(a)$ where a is a given natural number. In the fourth section, such an information is given in terms of the quantities

$$e_\ell(k, a) := \left| \{ b \in \mathcal{P}_T(a) : k \text{ times } T_0, \ell \text{ times } T_1 \} \right|$$

counting the number of possible paths terminating at $a \in \mathbb{N}$ with precisely k edges T_0 and ℓ edges T_1. According to the encoding described above, these paths are precisley those which are encoded by non-negative integer vectors $(s_0, \ldots, s_\ell)$ satisfying $s_0 + \cdots + s_\ell = k$. We come to a theorem estimating the number of elements in a portion

$$\mathcal{P}_T^\nu(a) = \left\{ x \in \mathcal{P}_T(a) : 2^{\nu-1} a < x < 2^\nu a \right\},$$

where $\nu \in \mathbb{R}$ is arbitrarily given, by the series

$$|\mathcal{P}_T^\nu(a)| \geqslant \sum_{\ell=0}^\infty e_\ell \left(\left\lfloor \nu + \ell \log_2 \left(\frac{3}{2} \right) \right\rfloor, a \right).$$

Although this is formally written down as an infinite sum, it will be clear that only finitely many terms are nonzero. There are also some remarks about the error of this estimate.

The fifth section gives analoguous results for sets of odd predecessors, and for the *pruned* Collatz graph spanned by the vertices $a \in \mathbb{N}$ which are not divisible by 3, which has been introduced by J. C. Lagarias and A. Weiss [**LaW**].

In the last section, the concept of uniform bounds is introduced. Let $U \subset \mathbb{N}$ be an arbitrary subset such that to each $a \in U$, a subset $P(a) \subset \mathbb{N}$ is associated. Then a positive function φ on the positive real line is called *uniform lower bound* for the family of sets $\{P(a) : a \in U\}$, if

$$\liminf_{x \to \infty} \left(\inf_{a \in U} \frac{Z_{P(a)}(ax)}{x} \right) \geqslant 1 \,.$$

Here we take $U := \mathbb{N} \setminus 3\mathbb{N}$, and $P(a) \subset \mathcal{P}_T(a)$ an appropriate subset of the predecessor set $\mathcal{P}_T(a)$, e.g. the set of odd predecessors. Based on this notion, a thorough comparison of the concepts developed here and those introduced in the basic paper of R. E. Crandall [**Cra**] is given. It is shown that the tree-search method initiated by Crandall leads naturally to uniform lower bounds for the family of predecessor sets of numbers $a \in \mathbb{N} \setminus 3\mathbb{N}$. Some of the results of Crandall, and also of Sander [**San**], appear more natural in the setting of this chapter. The chapter concludes with a brief discussion of the 'minorant' vectors of D. Applegate and J. C. Lagarias [**AL1**].

1. Directed graphs and dynamical systems on $\mathbb{N}$

A function $g : \mathbb{N} \to \mathbb{N}$ gives rise to a dynamical system on $\mathbb{N}$ by associating to a given $n \in \mathbb{N}$ the trajectory $(n, f(n), f^2(n), \dots)$ generated by iterated application of f. It is worth to develop explicitly Collatz' idea to picture such a dynamical system by an infinite graph [**Col1, Col2**].

Directed graphs.   For definitness, we begin with explicit definitions of some fundamental notions of graph theory.

1.1. DEFINITION. A *directed graph* is a pair $\Gamma = (V, E)$ consisting of a set V of *vertices* and a set $E \subset V \times V$ of *directed edges*. A directed graph $\Gamma' = (V', E')$ is called a *subgraph* of Γ, if $V' \subset V$ and $E' \subset E \cap (V' \times V')$. A subgraph $\Gamma' = (V', E')$ of Γ is called the *full subgraph of* Γ *generated by* V', if $E' = E \cap (V' \times V')$.

A finite directed graph is usually given in a picture where the vertices are represented by points and the directed edges by arrows.

1.2. DEFINITION. Let $\Gamma = (V, E)$ be a directed graph, and fix two vertices $x, y \in V$.

(a) A finite sequence $\pi := (v_0, \dots, v_k) \subset V$ is called a (directed) *path from x to y*, if $v_0 = x$, $v_k = y$, and $(v_{j-1}, v_j) \in E$ for $j = 1, \dots, k$; in this case k is called *length* of the path π, in symbols $\ell(\pi) = k$, x is its *initial vertex*, in symbols $i(\pi) = x$, and y is its *terminal vertex*, in symbols $t(\pi) = y$. The set of all paths of the graph Γ is denoted by $\Pi(\Gamma)$.

(b) A path π is called a *path between x and y*, if π is either a path from x to y or a path from x to y.

(c) A path $(v_0, \ldots, v_k)$ is called a *cycle*, if $v_0 = v_k$ and $v_0 \neq v_j$ for $j = 1, \ldots, k-1$.

(d) If $\pi_1 = (v_0, \ldots, v_k)$ and $\pi_2 = (w_0, \ldots, w_\ell)$ are two paths in Γ with $t(\pi_1) = i(\pi_2)$, their *concatenation* is again a path defined by $\pi_1 \pi_2 := (v_0, \ldots, v_k, w_1, \ldots, w_\ell)$.

(e) A vertex v is called a *predecessor of y in* Γ, if there exists a path from v to y.

(f) A vertex t is called *terminal vertex in the graph* Γ, if, for each $v \in V$, there exists a path from v to t.

(g) The sequence $(v_0, \ldots, v_k)$ is called an *undirected path from $x = v_0$ to $y = v_k$*, if, for $j = 0, \ldots, k-1$, we have either $(v_j, v_{j+1}) \in E$ or $(v_{j+1}, v_j) \in E$ (or both).

Note that we shall use the word "path" without further qualification only if we want to refer to a *directed* path. If we want to refer to an undirected path, we have to state this explicitly.

1.3. REMARK. Each edge $e \in E$ of a graph $\Gamma = (V, E)$ is a path of length 1, whereas each vertex $v \in V$ can be considered as a path of length 0. Each path $\pi = (v_0, \ldots, v_k)$ of length $k \geqslant 0$ has a unique decomposition $\pi = e_1 \cdots e_k$ into k edges $e_j := (v_{j-1}, v_j)$, $j = 1, \ldots, k$. Moreover, length behaves additively under concatenation of paths, $\ell(\pi_1 \pi_2) = \ell(\pi_1) + \ell(\pi_2)$ whenever π_1, π_2 are two paths whose concatenation is defined.

The following notion is especially important to picture dynamical systems on discrete sets.

1.4. DEFINITION. Let $\Gamma = (V, E)$ be a directed graph, and let $a \in V$. A vertex $x \in V$ is said to be *weakly joined* to a, if there is an undirected path from x to a in Γ. The *weak component* of Γ containing the vertex a is defined to be the full subgraph $\widetilde{\Gamma}(a)$ generated by the set of vertices $x \in V$ which are weakly joined to a. The graph Γ is called *weakly connected*, if $\Gamma = \widetilde{\Gamma}(a)$.

Of course, the relation "x is weakly joined to a" is an equivalence relation on the set of vertices of a directed graph. So, for any given directed graph $\Gamma = (V, E)$, we have a natural decomposition of V into pairwise disjoint subsets V_α, $\alpha \in A$, such that the full subgraph generated by V_α is a weak component of Γ, for each $\alpha \in A$.

1.5. DEFINITION. A directed graph $\Gamma = (V, E)$ is called a *directed tree*, if Γ is weakly connected and $\Pi(\Gamma)$ contains no cycles.

1.6. REMARK. We shall have occasion to use the following characterization of directed trees. A directed graph $\Gamma = (V, E)$ is a directed tree, if, and only if, for every two vertices $x, y \in V$, there is at least one undirected path from x to y and at most one path between x and y. Note that this implies that a tree has at most one terminal vertex.

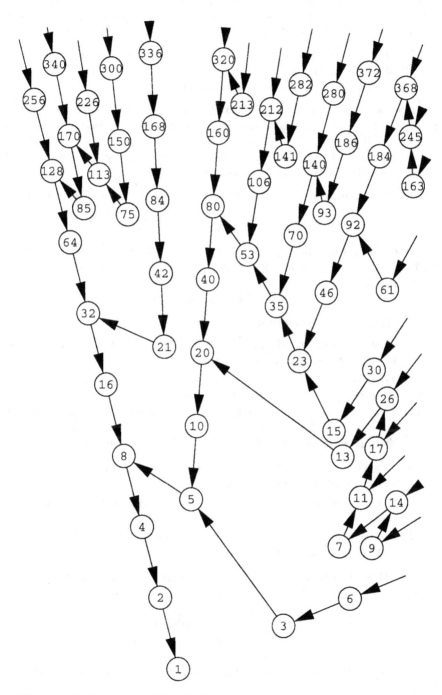

FIGURE 1. A portion of the Collatz graph of the $3n + 1$ function.

The Collatz graph.   We now discuss the relation between dynamical systems on $\mathbb{N}$ and directed graphs. Observe again that a dynamical system on $\mathbb{N}$ is just a function $f : \mathbb{N} \to \mathbb{N}$. Now we are ready to define the Collatz graph.

1.7. DEFINITION. Let $f : \mathbb{N} \to \mathbb{N}$ be an arbitrary integer function. The *Collatz graph* $\Gamma_f = (V_f, E_f)$ *of* f is defined by $V_f := \mathbb{N}$ as set of vertices and $E_f := \{(n, f(n)) : n \in \mathbb{N}\}$ as set of directed edges.

As we are—in these notes—especially concerned with the $3n + 1$ function T, the word *Collatz graph* without further specification will always refer to the Collatz graph Γ_T of T.

We proceed to discuss the relation between dynamical systems on $\mathbb{N}$ and the graphs derived from the associated integer functions.

1.8. DEFINITION. Let $f : \mathbb{N} \to \mathbb{N}$ be given, and let $a \in \mathbb{N}$. The sequence

$$\mathcal{T}_f(a) := \left(f^k(a)\right)_{k \geqslant 0} = (a, f(a), f \circ f(a), \dots)$$

of f-iterates of a is called $(f\text{-})$*trajectory* of a.   The number a is called *cyclic* (in Γ_f), if there is an integer $k > 0$ such that $f^k(a) = a$; otherwise it is called *non-cyclic*. The set

$$\mathcal{P}_f(a) := \{n \in \mathbb{N} : a \in \mathcal{T}_f(n)\}$$

is called $(f\text{-})$*predecessor set* of a.

1.9. REMARK. For an arbitrary function $f : \mathbb{N} \to \mathbb{N}$, there are two possibilities for the limiting behaviour of a trajectory:

(1) $\mathcal{T}_f(a)$ is bounded. Then the trajectory is eventually cyclic, i.e. there are integers $k_0, p > 0$ such that $f^k(a) = f^{k+p}(a)$ for $k \geqslant k_0$.

(2) $\mathcal{T}_f(a)$ is unbounded. Then $\mathcal{T}_f(a)$ can visit any bounded subset of $\mathbb{N}$ only finitely many times. In other words: For any threshold value $s \in \mathbb{N}$, there is an integer $k_s > 0$ such that $f^k(a) \geqslant s$ for $k \geqslant k_s$.

1.10. REMARK. For a subset $S \subset \mathbb{N}$, let $\langle S \rangle_f$ denote the full subgraph of Γ_f generated by the set of vertices S. Then:

(1) If $a \in V_f$ is non-cyclic, then $\langle \mathcal{P}_f(a) \rangle_f$ is a directed tree with terminal vertex a.

(2) If $a \in V_f$ is cyclic, then $\langle \mathcal{P}_f(a) \rangle_f$ consists of a cycle with directed trees attached to each of its vertices.

1.11. DEFINITION. Let $f : \mathbb{N} \to \mathbb{N}$ and $a \in \mathbb{N}$. The *domain of attraction* of the trajectory $\mathcal{T}_f(a)$ is defined by

$$\mathcal{A}_f(a) := \bigcup_{n \in \mathcal{T}_f(a)} \mathcal{P}_f(n) \,.$$

1.12. REMARK. The domain of attraction of a trajectory $\mathcal{T}_f(a)$ is also given by

$$\mathcal{A}_f(a) = \left\{n \in \mathbb{N} : \text{there exist integers } k, \ell \geqslant 0 \text{ with } f^k(n) = f^\ell(a)\right\} \,.$$

Moreover, the full subgraph of Γ_f generated by the set of vertices $\mathcal{A}_f(a)$ is just the weak component of a in Γ_f, in symbols $\widetilde{\Gamma}_f(a) = \langle \mathcal{A}_f(a) \rangle_f$. The graph theoretical structure of $\widetilde{\Gamma}_f(a)$ is related to the limiting behaviour or $\mathcal{T}_f(a)$ in the following way:

(1) If $\mathcal{T}_f(a)$ remains bounded, then $\widetilde{\Gamma}_f(a)$ consists of a cycle with a finite or infinite tree attached to each of the cycle's vertices, and each cycle-vertex is the terminal vertex of the tree attached to it.

(2) If $\mathcal{T}_f(a)$ is unbounded, then $\widetilde{\Gamma}_f(a)$ is an infinite tree without terminal vertex.

1.13. REMARK. Any weak component of Γ_f contains at most finitely many cyclic vertices.

The size of a subset of $\mathbb{N}$. What kind of information should we seek, if we want to know something about such a dynamical system? As each weak component of Γ_f is just the domain of attraction of one f-trajectory, a natural question would concern the "size" of weak components. More specific information concerning the "size" of the predecessor sets $\mathcal{P}_f(a) \subset \mathbb{N}$ would also be of interest.

What is needed is a good concept of "size" for subsets of $\mathbb{N}$. All possible information about a subset of $\mathbb{N}$ is encoded in the following functions.

1.14. DEFINITION. Let $C \subset \mathbb{N}$ be an arbitrary subset. The *counting function of C* is defined by

$$Z_C : \mathbb{N} \to \mathbb{N}_0, \qquad Z_C(x) := |\{y \in C : y \leqslant x\}|,$$

and the *weighted counting function of C* is given by

$$W_C : \mathbb{N} \to \mathbb{Q}, \qquad W_C(x) := \frac{|\{y \in C : y \leqslant x\}|}{x} = \frac{Z_C(x)}{x}.$$

In this setting, the "size" of a given subset $C \subset \mathbb{N}$ is encoded in the asymptotic behaviour, for large x, of its counting function Z_C. The asymptotic behaviour (as $x \to \infty$) of the weighted counting function represents, intuitively, the probability that a randomly chosen natural number is an element of C.

Because of its importance here, we use the following special notation for $3n+1$ predecessor counting functions.

1.15. NOTATION. Let $a \in \mathbb{N}$. The *predecessor counting function of a* is defined by

$$Z_a : \mathbb{N} \to \mathbb{N}_0, \qquad Z_a(x) := Z_{\mathcal{P}_T(a)} |\{y \in \mathcal{P}_T(a) : y \leqslant x\}|.$$

It is now clear that we can easily translate the $3n + 1$ conjecture into the different pictures given here in the following way:

1.16. COROLLARY. *For the $3n + 1$ function T, the following assertions are equivalent:*

(a) *For each $n \in \mathbb{N}$ there exists an integer $k \geqslant 0$ with $T^k(n) = 1$.*
(b) *$1 \in \mathcal{T}_T(n)$ for each $n \in \mathbb{N}$.*
(c) *$\mathcal{A}_T(1) = \mathcal{A}_T(2) = \mathbb{N}$.*
(d) *The weighted counting function of $\mathcal{P}_T(1)$ is constant 1.*
(e) *$Z_1(x) = x$ for all $x \in \mathbb{N}$.*
(f) *The Collatz graph of T is weakly connected.*

2. Encoding of predecessors by admissible vectors

This is not a text about general dynamical systems on discrete sets, nor on the general behaviour of iterations of integer functions. The topic which we discuss here is mainly the $3n + 1$ problem—which does not exclude the possibility that the methods and ideas used here may be of some value to treat a more general question.

But for the time being, let us restrict attention to the $3n + 1$ function

$$T : \mathbb{N} \to \mathbb{N}, \quad T(n) = \begin{cases} T_0(n) = n/2 & \text{if } n \text{ is even,} \\ T_1(n) = (3n + 1)/2 & \text{if } n \text{ is odd,} \end{cases}$$

and to the Collatz graph $\Gamma_T = (V_T, E_T)$ associated to it.

Encoding a path in the Collatz graph. Suppose we are given a path in the Collatz graph, i.e. a finite sequence of natural numbers $(v_0, \ldots, v_k) = \big(a, T(a), \ldots, T^k(a)\big)$, or

$$v_0 \xrightarrow{T} v_1 \xrightarrow{T} \cdots \xrightarrow{T} v_k \ .$$

As the $3n + 1$ function T has two branches, T_0 and T_1, some of the maps $T(v_{j-1}) = v_j$ are effected by T_0, and others by T_1. For example:

$$36 \xrightarrow{T_0} 18 \xrightarrow{T_0} 9 \xrightarrow{T_1} 14 \xrightarrow{T_0} 7 \xrightarrow{T_1} 11 \xrightarrow{T_1} 17 \xrightarrow{T_1} 26$$

Writing a 0 for each occurence of T_0 and a 1 for each occurence of T_1, such a path gives rise to a 0-1-vector, e.g. $(0, 0, 1, 0, 1, 1, 1)$. The 0-1-vector so obtained has been called *encoding vector* by Terras [**Ter1**] or *parity vector* by Lagarias [**Lag1**], referring to the infinite *parity sequence* of Everett [**Eve**].

2.1. DEFINITION. Let $\pi = (v_0, \ldots, v_k)$ be a path in Γ_T. The 0-1-vector $(x_1, \ldots, x_k)$ defined by $v_j = T_{x_j}(v_{j-1})$ for $j = 1, \ldots, k$ is called *parity vector* of the path π.

We shall use here a slightly more complicated *encoding vector* which consists of finitely many non-negative integers and which can be constructed out of the parity vector in the following way: Represent each 1 in the parity vector by a comma, and write at the beginning, at the end, and between two neighbouring commas just the number of 0's occuring at the beginning, at the end, and between the two 1's which are represented by the two neighbouring commas, respectively. For instance, we map

$$(0,0,1,0,1,1,1) \quad \longmapsto \quad (2,1,0,0,0) .$$

Concatenation of integer vectors. We now start to give a machinery leading to a strict formalization of the procedure described above.

2.2. DEFINITION. Denote by $\mathcal{F}$ the set of finite sequences of non-negative integers,

$$\mathcal{F} := \{ s = (s_0, \ldots, s_k) : \ k \in \mathbb{N}_0, \text{ and } s_j \in \mathbb{N}_0 \text{ for } j = 0, \ldots, k \} .$$

The elements $s \in \mathcal{F}$ are called *(non-negative) integer vectors*, or just *feasible vectors*. For a feasible vector $s = (s_0, \ldots, s_k)$, we define

$$\begin{array}{lll} \text{the } \textit{length}: & \ell(s) := k & (\text{not } k+1), \\ \text{the } \textit{absolute value}: & |s| := s_0 + \cdots + s_k, \\ \text{and the } \textit{norm}: & \|s\| := |s| + \ell(s). \end{array}$$

Our aim is to construct a map

$$\sigma : \Pi(\Gamma_T) \longrightarrow \mathcal{F}$$

from the set $\Pi(\Gamma_T)$ of paths in the Collatz graph into $\mathcal{F}$ in such a way that concatenation of paths is reflected by an appropriate concatenation of integer vectors. The crucial point in this construction is the definition of the latter concatenation, thereby formalizing the ideas described in the previous subsection.

2.3. DEFINITION. Let $u = (s_0, \ldots, s_k), t = (t_0, \ldots, t_\ell) \in \mathcal{F}$. Then their *concatenation* is defined by

$$u \cdot t := (s_0, \ldots, s_{k-1}, s_k + t_0, t_1, \ldots, t_\ell) \in \mathcal{F}.$$

Conversely, an equation $s = u \cdot t$ in $\mathcal{F}$ is called *decomposition* of s into the pair (u, t). If $s = u \cdot t$ is a decomposition of $s \in \mathcal{F}$, then we call u an *initial part* and t a *terminal part* of s.

2.4. LEMMA. *Let $r, s, t \in \mathcal{F}$. Then:*

(a) *The vector $(0) \in \mathcal{F}$ is neutral for concatenation, $(0) \cdot r = r \cdot (0) = r$.*
(b) *Concatenation is associative, $(r \cdot s) \cdot t = r \cdot (s \cdot t)$.*
(c) *Left and right cancellation are valid:*

$$t \cdot r = t \cdot s \ \Rightarrow \ r = s, \qquad t \cdot r = s \cdot r \ \Rightarrow \ t = s.$$

(d) *Length, absolute value, and norm, are additive:*

$$\ell(s \cdot t) = \ell(s) + \ell(t), \qquad |s \cdot t| = |s| + |t|, \qquad \|s \cdot t\| = \|s\| + \|t\|.$$

The proofs of the assertions of this lemma are immediate.

2.5. LEMMA. *Let $s \in \mathcal{F}$. For any integer k with $0 \leqslant k \leqslant \|s\|$, there is a unique terminal part $\tau_k(s)$ of s with $\|\tau_k(s)\| = k$. Moreover, we have for each $k \in \{1, \ldots, \|s\|\}$ either $\tau_k(s) = (1) \cdot \tau_{k-1}(s)$ or $\tau_k(s) = (0,0) \cdot \tau_{k-1}(s)$.*

PROOF. By left and right cancellation, and because $(0) \in \mathcal{F}$ is neutral for concatenation, the claim is true for $k = 0$ and $k = \|s\|$.

If we write $s = (s_0, \ldots, s_\ell)$, we infer from definition 2.3:

$$s_\ell \neq 0 \iff s = (s_0, \ldots, s_{\ell-1}, s_\ell - 1) \cdot (1) \,,$$
$$s_\ell = 0 \iff s = (s_0, \ldots, s_{\ell-1}) \cdot (0,0) \,,$$
$$s_0 \neq 0 \iff s = (1) \cdot (s_0 - 1, s_1, \ldots, s_\ell) \,,$$
$$s_0 = 0 \iff s = (0,0) \cdot (s_1, \ldots, s_\ell) \,,$$

which proves the claim for $k = 1$ and $k = \|s\| - 1$.

Now suppose the claim is true for some $k < \|s\|$, and let $s = u \cdot \tau_k(s)$ be the unique decomposition of s with $\|\tau_k(s)\| = k$. Again, there is a unique decomposition $u = u' \cdot t'$ with $\|t'\| = 1$. Now associativity of concatenation of integer vectors gives $s = u' \cdot (t' \cdot t)$, proving that $\tau_{k+1}(s) := t' \cdot \tau_k(s)$ is a terminal part of s with $\|\tau_{k+1}(s)\| = k + 1$.

Uniqueness essentially follows from the same argument, as each terminal part with norm $k + 1$ gives rise, in a unique way, to a terminal part with norm k. $\quad\square$

2.6. THEOREM. *There is a unique map $\sigma : \Pi(\Gamma_T) \to \mathcal{F}$ satisfying:*

    (i) *If (v, w) is an edge in Γ_T and $w = T_0(v)$, then $\sigma(v, w) = (1)$.*

    (ii) *If (v, w) is an edge in Γ_T and $w = T_1(v)$, then $\sigma(v, w) = (0,0)$.*

    (iii) *If $\pi_1, \pi_2 \in \Pi(\Gamma_T)$ such that the concatenation $\pi_1\pi_2$ is defined, then $\sigma(\pi_1\pi_2) = \sigma(\pi_1) \cdot \sigma(\pi_2)$.*

PROOF. Existence and uniqueness follow from the fact that each path $\pi \in \Pi(\Gamma_T)$ has a unique decomposition into edges, and from associativity of concatenation of integer vectors. $\quad\square$

2.7. DEFINITION. For a path $\pi \in \Pi(\Gamma_T)$, the integer vector $\sigma(\pi) \in \mathcal{F}$ is called *encoding vector* of π.

Note that we have, by construction, the following facts.

2.8. COROLLARY. *If $\pi \in \Pi(\Gamma_T)$ has length k, then $\|\sigma(\pi)\| = k$. Moreover, if we take the unique decomposition $\pi = e_1 \ldots e_k$ of π into edges, $|\sigma(\pi)|$ is the number of edges arising from T_0, whereas $\ell(\sigma(\pi))$ is the number of edges arising from T_1.*

All what follows in this chapter can be viewed as a study of this map σ, aiming to uncover structures and regularities in the Collatz graph.

Tracing back integer vectors in the rationals.   We now return to the given facts that the vertices of the Collatz graph are natural numbers and that it is possible to follow a path computationally both forward and backward. With our construction of the encoding vector of a path in mind, we introduce the following operators.

2.9. DEFINITION. Let $s = (s_0, \ldots, s_k) \in \mathcal{F}$. The *back-tracing operator* v_s : $\mathbb{Q} \to \mathbb{Q}$ is given by

$$v_s := v_+^{s_0} \circ \left(v_- \circ v_+^{s_1} \right) \circ \cdots \circ \left(v_- \circ v_+^{s_k} \right),$$

where

$$v_+(q) := 2q, \quad v_-(q) := \tfrac{1}{3}(2q - 1) \qquad \text{for } q \in \mathbb{Q}.$$

Note that v_+ and v_- are precisely the inverse operators on the rationals of the two branches T_0 and T_1 of the $3n + 1$ function T, respectively. It is immediate that the back-tracing operators are well-behaved w.r.t. concatenation.

2.10. COROLLARY. *If* $s, t \in \mathcal{F}$, *then* $v_{s \cdot t} = v_s \circ v_t$.

The following lemma justifies the expression *back-tracing operator*.

2.11. LEMMA. *Let* $b \in V_T$, $k \in \mathbb{N}_0$, *and* $\pi_k(b) := \left(b, T(b), \ldots, T^k(b) \right)$ *the path in* Γ_T *with initial vertex* b *and length* k. *Then* $v_{\sigma(\pi_k(b))} \left(T^k(b) \right) = b$.

PROOF. (Induction on k.) The claim is clear for $k = 0$. Let $k = 1$. Then there are two cases: If b is even, then $T(b) = T_0(b)$ and $\sigma(\pi_1(b)) = (1)$, whence

$$v_{\sigma(\pi_1(b))}(T(b)) = v_+(T_0(b)) = b.$$

If b is odd, then $T(b) = T_1(b)$ and $\sigma(\pi_1(b)) = (0, 0)$, whence

$$v_{\sigma(\pi_1(b))}(T(b)) = v_-(T_1(b)) = b.$$

Now let $k > 1$. Write $\pi_k(b)$ as a concatenation $\pi_k(b) = \pi_1(b)\pi_{k-1}(T(b))$. We derive using theorem 2.6, corollary 2.10, and the induction hypothesis

$$v_{\sigma(\pi_k(b))} \left(T^k(b) \right) = v_{\sigma(\pi_1(b)) \cdot \sigma(\pi_{k-1}(T(b)))} \left(T^{k-1} \circ T(b) \right)$$
$$= v_{\sigma(\pi_1(b))} \left(v_{\sigma(\pi_{k-1}(T(b)))} \left(T^{k-1}(T(b)) \right) \right) = v_{\sigma(\pi_1(b))}(T(b)) = b,$$

which completes the proof. $\square$

In order to evaluate the back-tracing operators on rational numbers, we use the following quantities.

2.12. DEFINITION. For $s = (s_0, \ldots, s_k) \in \mathcal{F}$, we define

$$c(s) := \frac{2^{\|s\|}}{3^{\ell(s)}} \qquad \text{and} \qquad r(s) := \sum_{j=0}^{k-1} \frac{2^{j+s_0+\cdots+s_j}}{3^{j+1}},$$

and call $c(s)$ the *(backward) coefficient* and $r(s)$ the *(backward) remainder* of s.

2.13. LEMMA. *Let $q \in \mathbb{Q}$ and $s \in \mathcal{F}$. Then $v_s(q) = c(s)\, q - r(s)$.*

PROOF. (Induction on $k = \ell(s)$.) If $k = 0$, that is $s = (s_0)$, then $v_s(q) = 2^{s_0} q = c(s)\, q$; the sum in the formula for $r(s)$ is empty, hence $r(s) = 0$.

Now let $s = (s_0, \ldots, s_k)$ and $s' = (s_0, \ldots, s_{k-1})$ such that $s = s' \cdot (0, s_k)$. Then, using corollary 2.10 and the induction hypothesis,

$$v_s(q) = v_{s'} \circ v_{(0,s_k)}(q) = v_{s'} \left(\tfrac{2}{3} \cdot 2^{s_k} q - \tfrac{1}{3} \right)$$

$$= c(s') \left(\frac{2^{1+s_k}}{3} q - \frac{1}{3} \right) - r(s') = \frac{2^{\|s'\|+1+s_k}}{3^{\ell(s')+1}} q - \left(r(s') + \frac{2^{\|s'\|}}{3^{\ell(s')+1}} \right).$$

Observing that $\|s\| = \|s'\| + 1 + s_k$ and $\ell(s) = \ell(s') + 1$, and evaluating

$$r(s') + \frac{2^{\|s'\|}}{3^{\ell(s')+1}} = \sum_{j=0}^{k-2} \frac{2^{j+s_0+\cdots+s_j}}{3^{j+1}} + \frac{2^{k-1+s_0+\cdots+s_{k-1}}}{3^k} = r(s),$$

the result follows. $\square$

Lemma 2.13 is just a reversed, and more explicit, form of proposition 4 in [BöS].

2.14. COROLLARY. *Let $s, t \in \mathcal{F}$. Then coefficient and remainder fulfill* [*]

$$\begin{pmatrix} c(s \cdot t) & r(s \cdot t) \\ 0 & 1 \end{pmatrix} = \begin{pmatrix} c(s) & r(s) \\ 0 & 1 \end{pmatrix} \cdot \begin{pmatrix} c(t) & r(t) \\ 0 & 1 \end{pmatrix}.$$

PROOF. The claim follows by corollary 2.10 and the computation

$$v_s \circ v_t(q) = c(s)(c(t)\, q - r(t)) - r(s) = c(s)c(t)\, q - (c(s)r(t) + r(s)).   \square$$

Admissible integer vectors.   Let us return to the Collatz graph $\Gamma_T = (V_T, E_T)$, and let us fix a vertex $a \in V_T = \mathbb{N}$. Given an arbitrary feasible vector $s \in \mathcal{F}$, it is of course not generally true that the back-tracing operator v_s maps a onto another vertex $b \in V_T$. Since $V_T = \mathbb{N}$, this can only be true if $b = v_s(a) \in \mathbb{N}$.

2.15. DEFINITION. Let $a \in \mathbb{N}$ be given. We call $s \in \mathcal{F}$ *admissible* w.r.t. a, if $v_s(a) \in \mathbb{N}$, and use the notation $\mathcal{E}(a) := \{ s \in \mathcal{F} : v_s(a) \in \mathbb{N} \}$ for the set of admissible vectors w.r.t. a.

Our next aim is to show that each vector which is admissible w.r.t. $a \in V_T$ is in fact the encoding vector of a path in Γ_T with terminal vertex a. The following lemma gives some equivalent characterizations of admissible vectors.

---

[*] I owe this matrix representation to an anonymous referee.

2.16. LEMMA. *Let $s \in \mathcal{F}$ and $a \in \mathbb{N}$. Then the following assertions are equivalent:*

(a) $v_s(a) \in \mathbb{N}$.
(b) $v_s(a) \in \mathbb{Z}$.
(c) *For each terminal part t of s, we have $v_t(a) \in \mathbb{Z}$.*
(d) *For each terminal part t of s, we have $v_t(a) \in \mathbb{N}$.*

PROOF. (a) $\Rightarrow$ (b) is obvious.

(b) $\Rightarrow$ (c): Suppose $v_s(a) \in \mathbb{N}$ and choose a decomposition $s = r \cdot t$ in $\mathcal{F}$. Then lemma 2.13 implies

$$v_t(a) \in \left(\tfrac{1}{3}\right)^{\ell(t)} \mathbb{Z}.$$

On the other hand, corollary 2.10 and lemma 2.13 yield

$$v_s(a) = v_u(v_t(a)) = c(u) \, v_t(a) - r(u).$$

Hence

$$v_t(a) = \frac{3^{\ell(u)}}{2^{\|u\|}} \left(v_s(a) + r(u)\right) \in \left(\tfrac{1}{2}\right)^{\|u\|} \mathbb{Z},$$

because $3^{\ell(u)} r(u)$ is an integer. Combining these two gives

$$v_t(a) \in \left(\tfrac{1}{2}\right)^{\|u\|} \mathbb{Z} \cap \left(\tfrac{1}{3}\right)^{\ell(t)} \mathbb{Z} = \mathbb{Z},$$

where the last equality is due to the fact that $2^{\|u\|}$ and $3^{\ell(t)}$ are coprime.

For (c) $\Rightarrow$ (d), we assume that $v_t(a) \in \mathbb{Z}$ for any terminal part t of s, and it remains to show that $v_t(a) \geqslant 1$. We do this by induction on $k = \|t\|$.

For $k = 0$, we have $t = (0)$, and $v_t(a) = a \geqslant 1$.

By lemma 2.5, we know that there is a unique terminal part $\tau_k(s)$ of s with $\|\tau_k(s)\| = k$, for each $k \in \{1, \dots, \|s\|\}$, satisfying

either $\tau_k(s) = (1) \cdot \tau_{k-1}(s)$ or $\tau_k(s) = (0,0) \cdot \tau_{k-1}(s)$.

Assuming inductively that $v_{\tau_{k-1}(s)}(a) \geqslant 1$, we have in the first case

$$v_{\tau_k(s)}(a) = 2 \, v_{\tau_{k-1}(s)}(a) \geqslant 1,$$

and in the second case

$$v_{\tau_k(s)}(a) = \tfrac{1}{3}(2 \, v_{\tau_{k-1}(s)}(a) - 1) \geqslant \tfrac{1}{3},$$

which implies $v_{\tau_k(s)}(a) \geqslant 1$ because $v_{\tau_k(s)}(a) \in \mathbb{Z}$.

(d) $\Rightarrow$ (a) is immediate. $\square$

2.17. LEMMA. *Let $a \in \mathbb{N}$ and $s \in \mathcal{E}(a)$. Then $T^{\|s\|}(v_s(a)) = a$.*

PROOF. (Induction on $k = \|s\|$.) The lemma is easily checked for $\|s\| \leqslant 1$, because there are only three feasible vectors $s \in \mathcal{F}$ with $\|s\| \leqslant 1$, namely (0), (1), and $(0,0)$.

Now let $s \in \mathcal{E}(a)$ with $\|s\| = k > 1$, and assume that $T^{\|t\|}(v_t(b)) = b$ for any $b \in \mathbb{N}$ and any $t \in \mathcal{E}(b)$ with $\|t\| < k$. then, according to lemma 2.5, there is a unique decomposition $s = u \cdot t$ with both $\|u\| < k$ and $\|t\| < k$. By lemma 2.16, we have $v_t(a) \in \mathbb{N}$, which means, by definition 2.15, $u \in \mathcal{E}(v_t(a))$. Now the induction hypothesis gives

$$T^{\|s\|}(v_s(a)) = T^{\|t\|+\|u\|}(v_u(v_t(a))) = T^{\|t\|}(v_t(a)) = a ,$$

which completes the proof.   □

2.18. THEOREM. *Let $\Pi^a(\Gamma_T)$ denote the set of paths in the Collatz graph with terminal vertex $a \in V_T$. Then the map σ of theorem 2.6 establishs a one-one-correspondence*

$$\Pi^a(\Gamma_T) \longleftrightarrow \mathcal{E}(a) .$$

PROOF. Let $\pi = (b, v_1, \ldots, v_{k-1}, a) \in \Pi^a(\Gamma_T)$ be a path of length k. Then $\pi = \pi_k(a)$ in the notation of lemma 2.11, and we infer from that lemma that

$$v_{\sigma(\pi)}(a) = v_{\sigma(\pi)}(T^k(b)) = b \in \mathbb{N} ,$$

which means $\sigma(\pi) \in \mathcal{E}(a)$.

On the other hand, each vector $s \in \mathcal{E}(a)$ gives rise to a path

$$p(s) = \Big(v_s(a), T(v_s(a)), \ldots, T^{\|s\|}(v_s(a))\Big) ,$$

and we know by lemma 2.17 that $t(p(s)) = T^{\|s\|}(v_s(a)) = a$, which means $p(s) \in \Pi^a(\Gamma_T)$.

Now the theorem follows from the relations $\sigma(p(s)) = s$ and $p(\sigma(\pi)) = \pi$.   □

2.19. THEOREM. *If $a \in V_T$ is non-cyclic, then the map $\mathcal{E}(a) \to \mathcal{P}_T(a)$, $s \mapsto v_s(a)$, is a bijection.*

PROOF. The map in question arises as a composition of the map $p : \mathcal{E}(a) \to \Pi^a(\Gamma_T)$ which has been considered in the proof of the previous theorem, and the map $i : \Pi^a(\Gamma_T) \to V_T$ associating to each path π its initial vertex $i(\pi)$.

The first map p is a bijection, by theorem 2.18. In addition, we know by remark 1.10 that $\mathcal{P}_T(a)$ generates a directed tree, if $a \in V_T$ is non-cyclic. But, by remark 1.6, there is at most one path between to vertices of a directed tree. Hence, the second map i is also a bijection, and the theorem is proved.   □

3. Some properties of admissible vectors

The sets $\mathcal{E}(a)$ (defined in definition 2.15) of non-negative integer vectors which are admissible w.r.t. some given $a \in \mathbb{N}$ have some interesting intrinsic structure, which is reflected by certain properties of the Collatz graph. Basically, this structure arises from the fact that 2 is a primitive root modulo 3^{ℓ}, for each $\ell \in \mathbb{N}$, i.e. the powers 2^k, $k \in \mathbb{N}$, run through all residue classes modulo 3^{ℓ} whose elements are not divisible by 3.

Recognizing admissible vectors. It has been proved by Terras [**Ter1**] and Everett [**Eve**] that, for any given integer $k > 0$, the function $n \mapsto E_k(n)$ mapping a natural number n onto its parity vector (see definition 2.1) is both surjective onto the set of 0-1-vectors of length k and periodic with period 2^k. This means that any given 0-1-vector $(x_0, \ldots, x_k)$ is realized as parity vector $E_k(n)$ exactly by the natural numbers n contained in one residue class $a \pmod{2^k}$. There is a similar statement about the encoding vector discussed here.

3.1. LEMMA. *Let $s \in \mathcal{F}$ be given. Then there is exactly one residue class a $(\mathrm{mod}\ 3^{\ell(s)})$ such that*

$$s \in \mathcal{E}(b) \quad \Longleftrightarrow \quad b \equiv a \mod 3^{\ell(s)} .$$

PROOF. By definition 2.15 and lemma 2.16, we have $s \in \mathcal{E}(b) \Leftrightarrow v_s(b) \in \mathbb{Z}$. Lemma 2.13 gives

$$v_s(b) = c(s)\, b - r(s) = \frac{1}{3^{\ell(s)}} \left(2^{\|s\|} b - 3^{\ell(s)} r(s) \right) ,$$

where $d := 3^{\ell(s)} r(s) \in \mathbb{N}$. Hence

$$s \in \mathcal{E}(b) \quad \Longleftrightarrow \quad 2^{\|s\|} b \equiv d \mod 3^{\ell(s)} .$$

As $2^{\|s\|}$ and $3^{\ell(s)}$ are coprime, this congruence is solved by a unique residue class $a \pmod{3^{\ell(s)}}$. $\square$

This lemma justifies the term "feasible" for the non-negative integer vectors $s \in \mathcal{F}$.

Extending admissible vectors. We now start to investigate the possibilities to extend a given feasible vector in such a way that the extended vector retains (or gains) the property of being admissible w.r.t. some given $b \in \mathbb{N}$.

In principle, a feasible vector $s \in \mathcal{F}$ may be extended in two different ways: either by taking $t \cdot s$ or by taking $s \cdot t$ as the extended vector. We give three lemmas on this subject.

3.2. LEMMA. *Let $a \in \mathbb{N}$ and $s \in \mathcal{E}(a)$ be given.*
(a) *We have $(k) \cdot s \in \mathcal{E}(a)$ for any $k \in \mathbb{N}_0$.*
(b) *If $a \not\equiv 0 \mod 3$, then there is a $b \in \mathbb{N}$ with $b \equiv a \mod 3^{\ell(s)}$ such that $(k,0) \cdot s \in \mathcal{E}(b)$ for any $k \in \mathbb{N}_0$.*

PROOF. (a) By definition 2.15, $v_s(a) \in \mathbb{N}$. Corollary 2.10 and definition 2.9 give

$$v_{(k) \cdot s}(a) = v_{(k)}(v_s(a)) = 2^k \, v_s(a) \in \mathbb{N} \quad \Longrightarrow \quad (k) \cdot s \in \mathcal{E}(a) \,.$$

(b) Because $(k, 0) \cdot s = (k) \cdot (0, 0) \cdot s$, it suffices to find $b \in \mathbb{N}$ such that $(0, 0) \cdot s \in \mathcal{E}(b)$. We have, for each $b \in \mathbb{N}$, according to definition 2.9,

$$v_{(0,0) \cdot s}(b) = v_-\big(v_s(b)\big) = \tfrac{1}{3}\big(2v_s(b) - 1\big) \,.$$

To make this an integer, we have to ensure $v_s(b) \equiv 2 \mod 3$. Let us put $b := a + 3^{\ell(s)} j$ and look for an appropriate $j \in \mathbb{Z}$. By lemma 2.13,

$$v_s(b) = c(s) \left(a + 3^{\ell(s)} j \right) - r(s) = v_s(a) + 2^{\|s\|} j \,.$$

Since $2^{\|s\|} \not\equiv 0 \mod 3$, we can solve the congruence $v_s(a) + 2^{\|s\|} j \equiv 2 \mod 3$ for j, which completes the proof. $\square$

3.3. LEMMA. Let $a, \ell \in \mathbb{N}$, $a \not\equiv 0 \mod 3$, $s = (s_0, \ldots, s_\ell) \in \mathcal{F}$, and let $k \in \mathbb{N}_0$. Then there is exactly one residue class b (mod $2 \cdot 3^\ell$) such that

$$(s_0, \ldots, s_\ell, k) = s \cdot (0, k) \in \mathcal{E}(a) \quad \Longleftrightarrow \quad k \equiv b \mod 2 \cdot 3^\ell \,.$$

PROOF. We know from lemma 3.1 that there is just one residue class c (mod $3^{\ell-1}$) such that, for $d \in \mathbb{N}$,

$$s \in \mathcal{E}(d) \quad \Longleftrightarrow \quad d \equiv c \mod 3^\ell \,.$$

Now the proof runs as follows:

$$s \cdot (0, k) \in \mathcal{E}(a) \quad \Longleftrightarrow \quad s \in \mathcal{E}\left(\frac{2^{k+1}a - 1}{3} \right) \quad \Longleftrightarrow \quad \frac{2^{k+1}a - 1}{3} \equiv c \mod 3^\ell$$

$$\Longleftrightarrow \quad 2^{k+1}a \equiv 3c + 1 \mod 3^{\ell+1} \,.$$

Because $3c + 1 \not\equiv 0 \mod 3$, because $a \not\equiv 0 \mod 3$, and because 2 generates the cyclic, multiplicative group of residue classes r (mod $3^{\ell+1}$) with $r \not\equiv 0 \mod 3$ (see, for instance, [Has], p. 81), we see that there is just one residue class b (mod $2 \cdot 3^\ell$) such that

$$2^{k+1}a \equiv 3c + 1 \mod 3^{\ell+1} \quad \Longleftrightarrow \quad k \equiv b \mod 2 \cdot 3^\ell \,. \quad \square$$

3.4. LEMMA. Let $a \in \mathbb{N}$. Then the following assertions are equivalent:
(a) $a \not\equiv 0 \mod 3$.
(b) There is an $s \in \mathcal{E}(a)$ with $\ell(s) = 1$.
(c) For each $\ell \in \mathbb{N}_0$, there is an $s \in \mathcal{E}(a)$ with length $\ell(s) = \ell$.

PROOF. (a) $\Rightarrow$ (b): If $a \equiv 1 \mod 3$, then $4a - 1 \equiv 0 \mod 3$, whence

$$v_{(0,1)}(a) = v_-(2a) = \frac{4a - 1}{3} \in \mathbb{N}.$$

This implies $(0, 1) \in \mathcal{E}(a)$, which is (b). If $a \equiv 2 \mod 3$, then $2a - 1 \equiv 0 \mod 3$, and we conclude $(0, 0) \in \mathcal{E}(a)$, yielding again (b).

(b) $\Rightarrow$ (c): By induction on ℓ, assume that there is some vector $s \in \mathcal{E}(a)$ with $\ell(s) = \ell \geqslant 1$. We have to show that there is some $t \in \mathcal{E}(a)$ with $\ell(t) = \ell(s) + 1$. We shall prove that there is some $u \in \mathcal{E}(a)$ satisfying both $\ell(u) = \ell(s)$ and $v_u(a) \not\equiv 0 \mod 3$. If we have achieved this, the step (a) $\Rightarrow$ (b) will show that there is vector $t' \in \mathcal{E}(v_u(a))$ with $\ell(t') = 1$, and $t := t' \cdot u$ will do.

Now let us construct u out of s. If $v_s(a) \not\equiv 0 \mod 3$, put $u := s$, and we are done. If $v_s(a) \equiv 0 \mod 3$, write $s = (s_0, s_1, \ldots, s_\ell) = (s_0, 0) \cdot t$ with $t = (s_1, \ldots, s_\ell)$ (remember $\ell \geqslant 1$). Then

$$v_s(a) = v_{(s_0,0)}(v_t(a)) = 2^{s_0} \frac{2v_t(a) - 1}{3} \equiv 3 \mod 3$$

implies $2v_t(a) \equiv 1 \mod 9$. This gives $2v_{(2) \cdot t}(a) = 2^3 v_t(a) \equiv 4 \mod 9$. Now put $u := (0, 2) \cdot t = (0, s_1 + 2, s_2, \ldots, s_\ell)$, and we are sure that

$$v_u(a) = \frac{2^3 v_t(a) - 1}{3} \not\equiv 0 \mod 3.$$

(c) $\Rightarrow$ (a): Let $s \in \mathcal{E}(a)$ have length 1, i. e. $s = (s_0, s_1)$. Then $v_s(a) = 2^{s_0}(2^{s_1+1}a - 1)/3 \in \mathbb{Z}$, which means $2^{s_1+1}a \equiv 1 \mod 3$. This implies (a). $\square$

3.5. COROLLARY. *Let $a \in \mathbb{N}$. Then the following assertions are equivalent:*

(a) $a \equiv 0 \mod 3$.
(b) $\mathcal{E}(a) = \{(k) : k \in \mathbb{N}_0\}$.
(c) $\mathcal{P}_T(a) = \{2^k a : k \in \mathbb{N}_0\}$.

PROOF. (a) $\Leftrightarrow$ (b) is another version of lemma 3.4, (a) $\Leftrightarrow$ (b).

(b) $\Rightarrow$ (c) is immediate.

(c) $\Rightarrow$ (b): Assume that $\mathcal{E}(a)$ contains a vector of length 1, say $s = (s_0, s_1)$. Then $v_s(a) \in \mathcal{P}_T(a)$ combines with (c) to prove

$$\frac{2^{s_1+1}a - 1}{3} = 2^k a \qquad \text{for some} \quad k \in \mathbb{N}_0.$$

This implies $(2^{s_1+1} - 3 \cdot 2^k)a = 1$, where each factor is a natural number. Consequently, $a = 1$. But we know very well that $\mathcal{P}_T(1) \neq \{2^k : k \in \mathbb{N}_0\}$, hence the assumption $(s_0, s_1) \in \mathcal{E}(a)$ is false. $\square$

Similar integer vectors. Motivated by the extension lemma 3.2, we now introduce an equivalence relation on the set $\mathcal{F}$ which behaves well w.r.t. the property of being admissible w.r.t. some $a \in \mathbb{N}$. This equivalence relation will be of some use in chapter IV.

3.6. DEFINITION. Let $s = (s_0, \ldots, s_k)$, $t = (t_0, \ldots, t_\ell) \in \mathcal{F}$. Put

$$s \simeq t \quad \Longleftrightarrow \quad \begin{cases} \ell(s) = \ell(t) & \text{and} \\ s_j \equiv t_j \mod 2 \cdot 3^{j-1} & \text{for each } j \in \{1, \ldots, \ell(s)\}. \end{cases}$$

We say that s and t are *similar*, if $s \simeq t$. The set $\{s\}^{\simeq} = \{t \in \mathcal{F} : t \simeq s\}$ is called the *similarity class of s*.

It is clear that this relation $s \simeq t$ is an equivalence relation on the set $\mathcal{F}$. Moreover, we now show that the similarity classes are not torn apart by the sets $\mathcal{E}(a)$.

3.7. LEMMA. *Let $a \in \mathbb{N}$, $s \in \mathcal{E}(a)$, and $t \in \mathcal{F}$ such that $s \simeq t$. Then $t \in \mathcal{E}(a)$.*

PROOF. (Induction on $k = \ell(s)$.) For $\ell(s) = 0$, there is nothing to prove, because any vector $t \in \mathcal{F}$ with $\ell(t) = 0$ satisfies both $t \simeq s$ and $t \in \mathcal{E}(a)$.

Now let $k = \ell(s) = \ell(t) \geqslant 1$, and let $s \simeq t$. Then we have unique decompositions

$$s = s' \cdot (0, s_k) \qquad \text{and} \qquad t = t' \cdot (0, t_k),$$

with $\ell(s') = \ell(t') = k - 1$ and the properties

$$s' \simeq t' \qquad \text{and} \qquad s_k \equiv t_k \mod 2 \cdot 3^{k-1}.$$

We conclude from $s \in \mathcal{E}(a)$ that $s' \in \mathcal{E}(v_{(0,s_k)}(a))$. Then the induction hypothesis gives $t' \in \mathcal{E}(v_{(0,s_k)}(a))$. But $s_k \equiv t_k \mod 2 \cdot 3^{k-1}$ implies

$$v_{(0,s_k)}(a) \equiv v_{(0,t_k)}(a) \mod 3^{k-1}.$$

By lemma 3.1, this means $t' \in \mathcal{E}(v_{(0,t_k)}(a))$, which immediately implies $t \in \mathcal{E}(a)$. $\square$

Recurrent patterns in the Collatz graph. Lemma 3.1 implies that every feasible vector $s \in \mathcal{F}$ is realized in the Collatz graph infinitely many times. Moreover, there are certain infinite subgraphs of the Collatz graph which occur infinitely many times. We are now going to state and prove this rigorously.

It is possible to interpret these results as infinite recurrence of certain subgraphs in the Collatz graph. To state this precisely, let us fix two numbers $a, k \in \mathbb{N}$, where a is to be considered as vertex of the Collatz graph $\Gamma_T = (V_T, E_T)$. The recurrent subgraphs are constructed using certain vertices in the predecessor sets $\mathcal{P}_T(a)$ (see definition 1.8). Put

$$\mathcal{P}_T^{(\#T_1 \leqslant k)}(a) := \left\{ b \in \mathcal{P}_T(a) \,\middle|\, \begin{array}{l} \text{the path } \pi \text{ from } b \text{ to } a \text{ contains} \\ \text{at most } k \text{ edges arising from } T_1. \end{array} \right\}.$$

According to remark 1.10, we know that, if a happens to be noncyclic, the full subgraph generated by this set of vertices,

$$\Gamma_T(a, k) := \left\langle \mathcal{P}_T^{(\#T_1 \leqslant k)}(a) \right\rangle_T,$$

is a directed tree with terminal vertex a.

3.8. LEMMA. *Fix $k \in \mathbb{N}$ and $a, b \in V_T$, and suppose that $b \not\equiv 0 \mod 3$ and that a is non-cyclic. Then $\langle \mathcal{P}_T(b) \rangle_T$ contains a subgraph which is isomorphic to $\Gamma_T(a, k)$.*

PROOF. Combining theorem 2.19 and corollary 2.8, and using that $a \in V_T$ is non-cyclic, we have a one-one-correspondence

$$\mathcal{P}_T^{(\#T_1 \leqslant k)}(a) \longleftrightarrow \{s \in \mathcal{E}(a) : \ell(s) \leqslant k\} \ .$$

If $a \equiv 0 \mod 3$, then we know by lemma 3.4 that

$$\mathcal{E}(a) = \{s \in \mathcal{F} : \ell(s) = 0\} \subset \mathcal{E}(c) \quad \text{for each } c \in \mathbb{N}.$$

Thus, in this case, $\mathcal{P}_T^{(\#T_1 \leqslant k)}(a) = \{2^m a : m \in \mathbb{N}_0\}$, and any non-cyclic predecessor c of b (which clearly exists by remark 1.13) gives rise to a subgraph

$$\langle \{2^m a : m \in \mathbb{N}_0\} \rangle_T \cong \Gamma_T(a, k) \ .$$

If $a \not\equiv 0 \mod 3$, we infer from lemma 3.1 that

$$\{s \in \mathcal{E}(a) : \ell(s) \leqslant k\} = \{s \in \mathcal{E}(c) : \ell(s) \leqslant k\} \quad \Longleftrightarrow \quad c \equiv a \mod 3^k \ .$$

If c is non-cyclic, the graph $\Gamma_T(c, k)$ has the property that, for any vertex $d \in \mathcal{P}_T^{(\#T_1 \leqslant k)}(c)$, there is a unique path in $\Gamma_T(c, k)$ from d to c. Hence, in this case the one-one-correspondences

$$\mathcal{P}_T^{(\#T_1 \leqslant k)}(a) \longleftrightarrow \{s \in \mathcal{E}(a) : \ell(s) \leqslant k\} \longleftrightarrow \mathcal{P}_T^{(\#T_1 \leqslant k)}(c)$$

induce an isomorphism of the subgraphs $\Gamma_T(a, k)$ and $\Gamma_T(c, k)$.

It remains to show that there is a vertex $c \in \mathcal{P}_T(b)$ which is both non-cyclic and satisfying $c \equiv a \mod 3^k$. For this, we remember again the fact that the powers of 2 run through all residue classes $r \pmod{3^k}$ whose members are not divisible by 3. Hence, using $b \not\equiv 0 \mod 3$, we see that there is a residue class $\gamma \pmod{2 \cdot 3^{k-1}}$ such that, for $\beta \in \mathbb{N}_0$,

$$2^\beta b \equiv a \mod 3^k \quad \Longleftrightarrow \quad \beta \equiv \gamma \mod 2 \cdot 3^{k-1} \ .$$

But it is clear that $\langle \mathcal{P}_T(b) \rangle_T$ is weakly connected (cf. remark 1.10). Hence it has at most finitely many cyclic vertices (by remark 1.13), and we can conclude that there is a number $\beta^* \in \mathbb{N}_0$ such that

$$c := 2^{\beta^*} b \equiv a \mod 3^k$$

is non-cyclic, which completes the proof. □

3.9. THEOREM. *Let $k \in \mathbb{N}$, and let $a \in V_T$ denote a non-cyclic vertex of the Collatz graph Γ_T. Then, for each weak component Γ' of Γ_T, there is an infinite set of pairwise disjoint subgraphs of Γ' each of which being isomorphic to $\Gamma_T(a, k)$.*

PROOF. Let $\Gamma' = (V', E')$ be a weak component of Γ_T (as defined in 1.4). We shall construct inductively a sequence of vertices $(x_m)_{m \in \mathbb{N}} \subset V'$ such that, for each $m \in \mathbb{N}$, three conditions are satisfied:

(i) $x_m \not\equiv 0 \mod 3$.
(ii) x_m is non-cyclic in Γ'.
(iii) $\mathcal{P}_T(x_m) \cap \mathcal{P}_T^{(\#T_1 \leqslant k)}(x_j) = \varnothing$ for $j = 1, \ldots, m - 1$.

Having at hand such a sequence $(x_m)_{m \in \mathbb{N}}$, the theorem follows from lemma 3.8.

To construct x_1, take any vertex $b \in V'$. As Γ' is a weak component, the $3n + 1$ trajectory $\left(T^j(b)\right)_{j \in \mathbb{N}_0}$ consists of vertices of V'. Let h be the highest number in $\mathbb{N}_0$ such that 2^h divides b. Then $c := T^h(b)$ is odd, and we have

$$T(c) = T_1(c) = \tfrac{1}{2}(3c + 1) \not\equiv 0 \mod 3 .$$

But we know by remark 1.13 that among the infinitely many vertices $2^\beta T(c)$, $\beta \in \mathbb{N}_0$, only finitely many can be cyclic. Thus, there is a number $\beta^* \in \mathbb{N}_0$ such that

$$x_1 := 2^{\beta^*} T(c)$$

satisfies conditions (i) and (ii). (For $m = 1$, (iii) is vacuous.)

Now suppose that $x_1, \ldots, x_m$ are already fixed, subject to conditions (i)–(iii). Using (i) for x_m, lemma 3.4 implies that there is a sequence $s \in \mathcal{E}(x_m)$ with $\ell(s) = k + 2$, say

$$s = (s_0, \ldots, s_{k+2}) = (s_0, 0) \cdot t \qquad \text{with} \quad t := (s_1, \ldots, s_{k+2}) .$$

Again lemma 3.4 implies that

$$x_{m+1} := v_t(x_m) \not\equiv 0 \mod 3 ,$$

hence condition (i) is satisfied for x_{m+1}. There is also no problem with (ii), because x_{m+1} is defined as a predecessor of the non-cyclic vertex x_m, whence x_{m+1} is also non-cyclic.

It remains to check condition (iii). Observe that $x_{m+1} \notin \mathcal{P}_T^{(\#T_1 \leqslant k)}(x_m)$, because x_m is non-cyclic, and because we have a path from x_{m+1} to x_m whose encoding vector t satisfies $\ell(t) = k + 1 > k$. From this we conclude that both

$$\mathcal{P}_T(x_{m+1}) \subset \mathcal{P}_T(x_m) \quad \text{and} \quad \mathcal{P}_T(x_{m+1}) \cap \mathcal{P}_T^{(\#T_1 \leqslant k)}(x_m) = \varnothing ,$$

which implies condition (iii). $\quad\square$

4. Counting functions and an estimate

Essentially, this section provides means to get rid of the Collatz graph in the study of the $3n + 1$ problem. We construct certain counting functions out of our analysis of the Collatz graph, and then we show that it is possible to obtain just these counting functions by an independent inductive process. Finally, an estimate designed to measure the "size" of $3n + 1$ predecessor sets is derived, which is based on that counting functions.

Counting functions for admissible vectors. For a given $a \in \mathbb{N}$, the admissible vectors w.r.t. a can be organized according to their length, or their absolute value, or both. We define counting functions reflecting this structure, and we prove some properties of them. Especially, we shall come across formulae which permit us to construct these counting functions inductively without reference to the Collatz graph.

4.1. DEFINITION. For $a \in \mathbb{N}$ and $k, \ell \in \mathbb{N}_0$, denote

$$\mathcal{E}_{\ell,k}(a) := \{s \in \mathcal{E}(a) : \ell(s) = \ell, |s| = k\} .$$

Now fix $\ell \in \mathbb{N}_0$. Then a *counting function for admissible vectors* is defined by

$$e_\ell : \mathbb{N}_0 \times \mathbb{N} \to \mathbb{N}_0, \qquad e_\ell(k, a) := |\mathcal{E}_{\ell,k}(a)| .$$

We record simple properties of these counting functions.

4.2. LEMMA. *For $a, b \in \mathbb{N}$ and $j, k, \ell \in \mathbb{N}_0$, we have*

(a) $e_0(k, a) = 1$.
(b) *If $k \geqslant j$, then $e_\ell(k, a) \geqslant e_\ell(j, a)$.*
(c) *If $\ell \geqslant 1$ and $a \equiv 0 \bmod 3$, then $e_\ell(k, a) = 0$.*
(d) *If $a \equiv b \bmod 3^\ell$, then $e_\ell(k, a) = e_\ell(k, b)$.*

PROOF. (a) is due to the elementary fact that $(k) \in \mathcal{F}$ is the unique feasible vector with length 0 and absolute value k, and that this vector (k) is admissible w.r.t. any $a \in \mathbb{N}$.

(b) follows from lemma 3.2 part (a), as $s \in \mathcal{E}_{\ell,j}(a)$ implies $(k - j) \cdot s \in \mathcal{E}_{\ell,k}(a)$ for any $a \in \mathbb{N}$.

(c) reformulates lemma 3.4.

(d) follows from lemma 3.1, as that lemma implies $\mathcal{E}_{\ell,k}(a) = \mathcal{E}_{\ell,k}(b)$ whenever $a \equiv b \bmod 3^\ell$. $\square$

One of the most interesting features of these counting functions is that it is possible to compute $e_{\ell+1}$ out of e_ℓ.

4.3. LEMMA. *Let $a \in \mathbb{N}$ and $k, \ell \in \mathbb{N}_0$, and put $e_\ell(j, q) := 0$ for $j \in \mathbb{N}_0$ and $q \in (\frac{1}{3}\mathbb{Z}) \setminus \mathbb{Z}$. Then*

$$e_{\ell+1}(k, a) = \sum_{j=0}^{k} e_\ell\left(k - j, \frac{2^{j+1}a - 1}{3}\right) .$$

PROOF. If we extend the notation of definition 4.1 by putting $\mathcal{E}_{\ell,j}(q) := \varnothing$ for $j \in \mathbb{N}_0$ and $q \in (\frac{1}{3}\mathbb{Z}) \setminus \mathbb{Z}$, then there is a one-one-correspondence

$$\mathcal{E}_{\ell+1,k}(a) \quad \longleftrightarrow \quad \bigcup_{j=0}^{k} \mathcal{E}_{\ell,k-j}\left(\frac{2^{j+1}a-1}{3}\right) ,$$

the union on the right hand side being a disjoint one. Indeed, an admissible vector

$$(s_0, \ldots, s_\ell, j) \in \mathcal{E}_{\ell+1,k}(a)$$

corresponds in a unique way to an admissible vector

$$(s_0, \ldots, s_\ell) \in \mathcal{E}_{\ell,k-j}\left(v_{(0,j)}(a)\right) = \mathcal{E}_{\ell,k-j}\left(\frac{2^{j+1}a-1}{3}\right) ,$$

and vice versa (cf. the proof of lemma 3.3). The lemma now follows by definition 4.1. □

4.4. COROLLARY. *The sequence of counting functions* $(e_\ell)_{\ell \in \mathbb{N}_0}$ *is uniquely determined by*

(i) $$e_0(k, a) = \begin{cases} 0 & \text{if } a \in \left(\bigcup_{\nu=1}^{\infty} \frac{1}{3^\nu} \mathbb{Z}\right) \setminus \mathbb{Z} , \\ 1 & \text{if } a \in \mathbb{Z} , \end{cases}$$

(ii) $$e_{\ell+1}(k, a) = \sum_{j=0}^{k} e_\ell\left(k-j, \frac{2^{j+1}a-1}{3}\right) \qquad \text{for } \ell \in \mathbb{N}_0 .$$

Bearing in mind part (d) of lemma 4.2, this remark allows us to define the counting functions on the product $\mathbb{N}_0 \times \mathbb{Z}_3$ where $\mathbb{Z}_3$ is the set 3-adic integers. We shall explicate and exploit this fact in chapter III.

Using the floor and ceiling functions, viz., for $x \in \mathbb{R}$,

$$\lfloor x \rfloor := \max\{n \in \mathbb{Z} : n \leqslant x\} , \qquad \lceil x \rceil := \min\{n \in \mathbb{Z} : n \geqslant x\} ,$$

respectively, we are now able to evaluate e_1.

4.5. COROLLARY. *Let* $a \in \mathbb{N}$ *and* $k \in \mathbb{N}_0$. *Then*

$$e_1(k, a) = \begin{cases} 0 & \text{if } a \equiv 0 \mod 3 , \\ \lfloor (k+1)/2 \rfloor & \text{if } a \equiv 1 \mod 3 , \\ \lceil (k+1)/2 \rceil & \text{if } a \equiv 2 \mod 3 . \end{cases}$$

PROOF. We have to use part (a) of lemma 4.2 and the formula of lemma 4.3:

$$e_1(k, a) = \sum_{j=0}^{k} e_0\left(k-j, \frac{2^{j+1}a-1}{3}\right) .$$

What is to do is to check whether the fraction is an integer. If $a \equiv 0 \mod 3$, then the denominator $2^{j+1}a - 1$ is never divisible by 3, and we obtain $e_1(k, a) = 0$.

Now let $a \equiv 1 \mod 3$. Then

$$\frac{2^{j+1}a - 1}{3} \in \mathbb{Z} \quad \Longleftrightarrow \quad j \equiv 1 \mod 2 \, ;$$

there are exactly $\lfloor (k+1)/2 \rfloor$ odd numbers j within the bounds $0 \leqslant j \leqslant k$. This gives the value of $e_1(k, a)$ in this case.

Similarly, $a \equiv 2 \mod 3$ implies that only the even j contribute to the sum computing e_1. As there are exactly $\lceil (k+1)/2 \rceil$ even numbers j within the bounds $0 \leqslant j \leqslant k$, the corollary follows. $\square$

Counting predecessors of given size. The counting functions e_ℓ introduced above can be used to obtain information about the "size" of predecessor sets $\mathcal{P}_T(a)$ as subsets of $\mathbb{N}$, provided that sufficently many things about the sequence $(e_\ell)_{\ell \in \mathbb{N}}$ are known. We are now showing how to do this.

Let us fix $a \in \mathbb{N}$. Then, clearly, the cardinalities of the sets

$$(4.1) \qquad \mathcal{P}_T^\nu(a) := \left\{ x \in \mathcal{P}_T(a) : 2^{\nu-1}a < x \leqslant 2^\nu a \right\}, \qquad \text{for } \nu \in \mathbb{R},$$

are of some relevance for estimating the counting functions of $\mathcal{P}_T(a) \subset \mathbb{N}$ given in definition 1.14. It is plain that, in case $a \equiv 0 \mod 3$, lemma 3.4 implies that $|\mathcal{P}_T^\nu(a)| = 1$ for each $\nu \in \mathbb{N}$. To handle other cases, we need more subtle considerations. To state the next lemma, recall definition 2.12 of *coefficient* $c(s)$ and *remainder* $r(s)$ of a feasible vector $s \in \mathcal{F}$, and put

$$(4.2) \qquad \mathcal{E}^\nu(a) := \left\{ s \in \mathcal{E}(a) : 2^{\nu-1} < c(s) \leqslant 2^\nu \right\} \, .$$

4.6. LEMMA. *Let $a \in \mathbb{N}$ be non-cyclic in Γ_T, and let $\nu \in \mathbb{R}$. Then there is a function $\beta_\nu : \mathcal{E}^\nu(a) \to \mathbb{N}_0$ such that the following map is a well-defined injection:*

$$\imath_\nu : \mathcal{E}^\nu(a) \to \mathcal{P}_T^\nu(a), \qquad \imath_\nu(s) := 2^{\beta_\nu(s)}v_s(a) \, .$$

PROOF. If $s \in \mathcal{E}^\nu(a)$ then, by lemma 2.13, $v_s(a) = c(s)\, a - r(s) \leqslant c(s)\, a$ since $r(s) \geqslant 0$. Thus, there is exactly one integer $\beta_\nu(s) \geqslant 0$ such that

$$2^{\nu-1}a < 2^{\beta_\nu(s)}v_s(a) \leqslant 2^\nu a \, .$$

To see injectivity of the map defined by $\imath_\nu(s) := 2^{\beta_\nu(s)}v_s(a)$, take a further vector $t \in \mathcal{E}^\nu(a)$, and suppose that $\imath_\nu(s) = \imath_\nu(t)$. Consequently,

$$v_{(\beta_\nu(s))\cdot s}(a) = 2^{\beta_\nu(s)}v_s(a) = 2^{\beta_\nu(t)}v_t(a) = v_{(\beta_\nu(t))\cdot t}(a) \, .$$

As a is noncyclic, this implies by theorem 2.19

$$(4.3) \qquad (\beta_\nu(s)) \cdot s = (\beta_\nu(t)) \cdot t \, .$$

Now it follows

$$2^{\beta_\nu(s)}c(s) = c\big((\beta_\nu(s)) \cdot s\big) = c\big((\beta_\nu(t)) \cdot t\big) = 2^{\beta_\nu(t)}c(t) \, ,$$

which gives

$$c(s) = 2^{\beta_\nu(t)-\beta_\nu(s)}c(t) \, .$$

Because $2^{\nu-1} < c(s) \leqslant 2^\nu$ and $2^{\nu-1} < c(t) \leqslant 2^\nu$, we infer $\beta_\nu(t) = \beta_\nu(s)$, which gives by left cancellation of (4.3) (which is valid by lemma 2.4) the desired relation $s = t$. $\square$

4.7. THEOREM. *Let $a \in \mathbb{N}$ be non-cyclic as a vertex of the Collatz graph of the $3n + 1$ function T, and let $\nu \in \mathbb{R}$. Then, with the constant $\lambda = \log_2(\frac{3}{2})$,*

$$|\mathcal{P}_T^\nu(a)| \geqslant \sum_{\ell=0}^{\infty} e_\ell(\lfloor \nu + \lambda\ell \rfloor, a) \ .$$

PROOF. Because a is assumed to be noncyclic, lemma 4.6 provides an injective map $\imath_\nu : \mathcal{E}^\nu(a) \to \mathcal{P}_T^\nu(a)$. Hence it suffices to estimate $|\mathcal{E}^\nu(a)|$.

$$
\begin{aligned}
s \in \mathcal{E}^\nu(a) &\iff& 2^{\nu-1} < c(s) \leq 2^\nu \\
&\iff& 2^{\nu-1} < 2^{|s|}\left(\tfrac{2}{3}\right)^{\ell(s)} \leq 2^\nu \\
&\iff& \nu - 1 < |s| - \log_2\left(\tfrac{3}{2}\right)\ell(s) \leq \nu \\
&\iff& \nu + \lambda\,\ell(s) - 1 < |s| \leq \nu + \lambda\,\ell(s) \\
&\iff& |s| = \lfloor \nu + \lambda\,\ell(s) \rfloor \\
&\iff& s \in \mathcal{E}_{\ell(s),\lfloor \nu + \lambda\,\ell(s) \rfloor}(a)
\end{aligned}
$$

Using, in addition, definition 4.1, we arrive at

$$|\mathcal{P}_T^\nu(a)| \geqslant |\mathcal{E}^\nu(a)| = \left| \bigcup_{\ell=0}^{\infty} \mathcal{E}_{\ell,\lfloor \nu + \lambda\ell \rfloor}(a) \right| = \sum_{\ell=0}^{\infty} e_\ell(\lfloor \nu + \lambda\ell \rfloor, a) \ . \quad \square$$

4.8. REMARK. For each particular $a \in \mathbb{N}$, only finitely many terms of the infinite sum in the estimate of this theorem are non-zero. On the other hand, lemma 3.1 shows that, for each given $\ell \in \mathbb{N}$ and $\nu \geqslant \lambda\ell$, there are numbers $a \in \mathbb{N}$ such that $e_\ell(\lfloor \nu + \lambda\ell \rfloor, a) \geqslant 1$. This means that it would be difficult to fix in advance the total number of non-zero terms in the infinite estimating sum of the preceding theorem.

There is one trivial exception: if $a \equiv 0 \mod 3$, then we know by lemma 4.2 that $e_\ell(k, a) = 0$ for $\ell \geqslant 1$, and $e_0(k, a) = 1$. On the other hand, in this case it is easily seen that $|\mathcal{P}_T^\nu(a)| = 1$ for each real number $\nu \geqslant 0$.

Let us reformulate this in terms of the counting function (cf. notation 1.15) of the predecessor set $\mathcal{P}_T(a)$.

4.9. COROLLARY. *Let $a \in \mathbb{N}$ be non-cyclic in Γ_T. Then, for each real $x > 0$,*

$$Z_a(x) \geqslant \sum_{n=0}^{\infty} \sum_{\ell=0}^{\infty} e_\ell\left(\left\lfloor \log_2\frac{x}{a} + \lambda\ell - n \right\rfloor, a\right) = \sum_{\ell=0}^{\infty} \sum_{k=0}^{\lfloor \log_2(x/a)+\lambda\ell \rfloor} e_\ell(k, a)$$

PROOF. For the first inequality, we infer from (4.1) that

$$\bigcup_{n=0}^{\infty} \mathcal{P}_T^{\log_2(x/a)-n}(a) = \{y \in \mathcal{P}_T(a) : y \leqslant x\} \ .$$

To obtain the inequality, apply theorem 4.7 to each individual term. The equation follows by rearranging terms and observing that $e_\ell(k, a) = 0$ for $k < 0$. $\quad \square$

For the estimate in this corollary, it is essential to assume that a is non-cyclic. The following lemma shows that, for any cyclic number a, there is an appropriate non-cyclic number $\tilde{a}$ which could serve in many contexts as a replacement for a.

4.10. LEMMA. *Let $a \in \mathbb{N}$ be cyclic in the Collatz graph Γ_T. Then there is a number $\tilde{a} \in \mathbb{N}$ with the following properties:*

(a) *$\tilde{a}$ is non-cyclic,*
(b) *$\mathcal{P}_T(\tilde{a}) \subset \mathcal{P}_T(a)$,*
(c) *$\tilde{a} \not\equiv 0 \mod 3$, and*
(d) *$a \leqslant \tilde{a} \leqslant 16\, a$.*

PROOF. If $a \equiv 0 \mod 3$, then we already know that $\mathcal{P}_T(a) = \{2^k a : k \in \mathbb{N}_0\}$; but this cannot be true as a is supposed to be cyclic. Hence $a \not\equiv 0 \mod 3$.

Now the strategy is to go through the remaining possible residue classes $a \mod 9$. For each $a \mod 9$ we calculate two predecessors $x, y \in \mathcal{P}_T(a)$ satisfying conditions (b), (c), and (d), which belong to different branches in $\mathcal{P}_T(a)$. This will, in each case separately, imply that either x or y is non-cyclic.

$a \equiv 1, 4 \mod 9$: Put $x := \frac{1}{3}(4a - 1)$ and $y := 4a$. Then $\tilde{a} = x$ and $\tilde{a} = y$ both satisfy (b), (c), and (d), and we have $T(x) = 2a = T(y)$ which implies that at least one of the two numbers x, y must be non-cyclic (otherwise the would have to belong to one cycle, which is impossible as neither is a predecessor of the other). Choosing $\tilde{a} \in \{x, y\}$ non-cyclic, we are ready.

$a \equiv 2, 8 \mod 9$: Put $x := \frac{2}{3}(2a - 1)$ and $y := 2a$. Again, $\tilde{a} = x$ and $\tilde{a} = y$ both satisfy (b), (c), and (d). Here we have $T^2(x) = a = T(y)$, implying that either x or y is non-cyclic.

$a \equiv 5 \mod 9$: Put $x := \frac{1}{3}(8a - 1)$ and $y := 8a$. Here we have $T(x) = 4a = T(y)$. (The choice $x := \frac{1}{3}(2a - 1)$ would not be possible, as then we would have $x \equiv 0 \mod 3$.)

$a \equiv 7 \mod 9$: Put $x := \frac{1}{3}(16a - 1)$ and $y := 16a$. Here $T(x) = 8a = T(y)$; as in the other cases, this implies that it is possible to choose a non-cyclic $\tilde{a} \in \{x, y\}$, which completes the proof. $\square$

The error of the estimate. What about the error in the estimate of theorem 4.7 ? The absolute error is just the lack of surjectivity of the map $\imath_\nu$ of lemma 4.6, i.e. the cardinality of $\mathcal{P}_T^\nu(a) \setminus \imath_\nu(\mathcal{E}^\nu(a))$. We do one further step aiming at an error estimate.

4.11. LEMMA. *Let $a \in \mathbb{N}$ be noncyclic in Γ_T, and let $\nu \in \mathbb{R}$. Then*

$$\mathcal{P}_T^\nu(a) \setminus \imath_\nu(\mathcal{E}^\nu(a)) = \left\{ v_t(a) \in \mathcal{P}_T^\nu(a) : t = (t_0, \dots, t_\ell) \in \mathcal{E}(a),\ c(t) > 2^{\nu + t_0} \right\}.$$

PROOF. The lemma will follow from the set equation

$$(4.4) \qquad \imath_\nu(\mathcal{E}^\nu(a)) = \left\{ v_t(a) \in \mathcal{P}_T^\nu(a) : t \in \mathcal{E}(a),\ c(t) \leqslant 2^{\nu + t_0} \right\}.$$

To prove the part "$\subset$", let $s \in \mathcal{E}^\nu(a)$. Then $t := \imath_\nu(s) = (\beta_\nu(s)) \cdot s$, where $\beta_\nu(s)$ is chosen such that $v_t(a) \in \mathcal{P}_T^\nu(a)$. From

$$t = (t_0, \dots, t_\ell) = (\beta_\nu(s) + s_0, s_1, \dots, s_\ell) \in \mathcal{F}$$

we infer that $\beta_\nu(s) \leqslant t_0$, which implies $c(t) = 2^{\beta_\nu(s)} c(s) \leqslant 2^{\nu + \beta_\nu(s)} \leqslant 2^{\nu + t_0}$.

It remains to show part "⊃" of (4.4). To this end, let $t = (t_0, \ldots, t_\ell) \in \mathcal{E}(a)$ satisfy both $v_t(a) \in \mathcal{P}_T^\nu(a)$ and $c(t) \leqslant 2^{\nu+t_0}$. The first condition gives

$$v_t(a) = c(t)a - r(t) > 2^{\nu-1}a,$$

which combines with the second one to yield

(4.5) $2^{\nu-1} < c(t) \leqslant 2^{\nu+t_0}.$

Now we have to find $s \in \mathcal{E}^\nu(a)$ such that $\imath_\nu(s) = (\beta_\nu(s)) \cdot s = t$. By (4.2), this s has to satisfy the condition

$$2^{\nu-1} < c(s) = 2^{-\beta_\nu(s)}c(t) \leqslant 2^\nu \quad \Longleftrightarrow \quad 2^{\beta_\nu(s)-1} < \frac{c(t)}{2^\nu} \leqslant 2^{\beta_\nu(s)}$$

$$\Longleftrightarrow \quad \beta_\nu(s) < \log_2 c(t) - \nu \leqslant \beta_\nu(s)$$

$$\Longleftrightarrow \quad \beta_\nu(s) = \lceil \log_2 c(t) \rceil$$

But (4.5) implies $\log_2 c(t) > \nu - 1$, whence $0 \leqslant \beta_\nu(s) \leqslant t_0$. □

4.12. REMARK. The formula of lemma 4.11 for the set $\mathcal{P}_T^\nu(a) \setminus \imath_\nu(\mathcal{E}^\nu(a))$ is not easy to evaluate. In fact, inserting the definition of $\mathcal{P}_T^\nu(a)$ into the right hand gives

$$\mathcal{P}_T^\nu(a) \setminus \imath_\nu(\mathcal{E}^\nu(a)) =$$
$$\left\{ v_t(a) : t \in \mathcal{E}(a), \ \max\left\{ 2^{\nu-1} + \frac{r(t)}{a}, 2^{\nu+t_0} \right\} < c(t) \leqslant 2^\nu + \frac{r(t)}{a} \right\}.$$

5. Some restricted predecessor sets

Partly for technical reasons and partly to increase our insight into the structure of the Collatz graph, we now describe some special counting functions and their relations to one another and to the $e_\ell(k, a)$ of definition 4.1. The first one counts odd predecessors, and the second one is associated to an interesting subgraph of the Collatz graph: the *pruned* Collatz graph. We add a discussion of a third function counting odd predecessors in the pruned Collatz graph; this will be of some value later.

The odd predecessors. It is occasionally good to restrict attention to odd vertices of the Collatz graph; let's see what happens if we do so. Let $a \in \mathbb{N}$, and let $s = (s_0, \ldots, s_\ell) \in \mathcal{E}(a)$. By definition 2.9, we immediately obtain

$$v_s(a) \equiv 1 \mod 2 \quad \Longleftrightarrow \quad s_0 = 0.$$

The next definition is reasonable because of this fact.

5.1. DEFINITION. For $a \in \mathbb{N}$, the *odd predecessor set* is given by

$$\mathcal{P}_T^o(a) := \{n \in \mathcal{P}_T(a) : n \equiv 1 \mod 2\} .$$

A feasible vector $s = (s_0, \ldots, s_\ell)$ is called *basic*, if $s_0 = 0$. The set of *basic admissible vectors* w.r.t. a is

$$\mathcal{E}^o(a) := \{(s_0, \ldots, s_\ell) \in \mathcal{E}(a) : s_0 = 0\} = \{s \in \mathcal{E}(a) : v_s(a) \text{ is odd.}\} .$$

For $\ell, k \in \mathbb{N}_0$, we use the notation (cf. 4.1)

$$\mathcal{E}_{\ell,k}^o(a) := \{s \in \mathcal{E}^o(a) : \ell(s) = \ell, |s| = k\} = \mathcal{E}_{\ell,k}(a) \cap \mathcal{E}^o(a) .$$

For fixed $\ell \in \mathbb{N}_0$, the *odd counting function* is

$$e_\ell^o : \mathbb{N}_0 \times \mathbb{N} \to \mathbb{N}_0, \qquad e_\ell^o(k, a) := \left| \mathcal{E}_{\ell,k}^o(a) \right| .$$

For explicitness, we note in passing what happens for $\ell = 0$:

$$\mathcal{E}_{0,k}^o(a) = \begin{cases} \{(0)\} & \text{if } k = 0 \text{ and } a \equiv 1 \mod 2 \\ \varnothing & \text{otherwise.} \end{cases}$$

The relation between the $e_\ell(k, a)$ and the $e_\ell^o(k, a)$ is readily derived.

5.2. LEMMA. *For $a \in \mathbb{N}$ and $\ell, k \in \mathbb{N}_0$, we have* $\quad e_\ell(k, a) = \sum_{j=0}^{k} e_\ell^o(j, a) .$

PROOF. The following map

$$\mathcal{E}_{\ell,k}(a) \longrightarrow \bigcup_{j=0}^{k} \mathcal{E}_{\ell,j}^o(a) , \qquad (s_0, s_1, \ldots, s_\ell) \mapsto (0, s_1, \ldots, s_\ell)$$

is clearly a bijection, as its inverse is given by putting $s_0 := k - (s_1 + \ldots + s_\ell)$. The lemma now follows from the definitions of the counting functions. $\square$

The next theorem gives an estimate for the counting function of the odd predecessor sets which is similar to that of corollary 4.9.

5.3. THEOREM. *Let $a \in \mathbb{N}$ be non-cyclic in the Collatz graph Γ_T, and put $\lambda = \log_2 3 - 1$. Then, for each real $x > 0$,*

$$Z_{\mathcal{P}_T^o(a)}(x) \geqslant \sum_{\ell=0}^{\infty} e_\ell \left(\left\lfloor \log_2 \frac{x}{a} + \lambda \ell \right\rfloor, a \right) .$$

PROOF. For any admissible vector $s \in \mathcal{E}_{\ell,j}(a)$, lemma 2.13 gives

$$v_s(a) = c(s) a - r(s) \leqslant c(s) a = \frac{2^{\ell+j} a}{3^\ell}$$

because $r(s) \geqslant 0$ (see definition 2.12). With the computation

$$\frac{2^{\ell+j}a}{3^\ell} \leqslant x \quad \Longleftrightarrow \quad 2^j \leqslant \frac{x}{a} \cdot \left(\frac{3}{2}\right)^\ell \quad \Longleftrightarrow \quad j \leqslant \log_2 \frac{x}{a} + \ell \cdot \log_2 \frac{3}{2}$$

we infer that $v_s(a) \leqslant x$ whenever $j \leqslant \log_2(x/a) + \lambda\ell$. If $a \in \mathbb{N}$ is non-cyclic, we know by theorem 2.19 that the map $s \mapsto v_s(a)$ is injective. Thus,

$$Z_{\mathcal{P}_T^o(a)} = |\{n \in \mathcal{P}_T^o(a) : n \leqslant x\}| \geqslant \left| \bigcup \left\{ \mathcal{E}_{\ell,j}^o(a) : j \leqslant \log_2 \frac{x}{a} + \ell\lambda \right\} \right|$$

$$= \sum_{\ell=0}^\infty \sum_{j=0}^{\lfloor \log_2(x/a)+\lambda\ell \rfloor} e_\ell^o(j,a) = \sum_{\ell=0}^\infty e_\ell\left(\left\lfloor \log_2 \frac{x}{a} + \lambda\ell \right\rfloor, a \right),$$

where we have used lemma 5.2 in the last step. $\square$

The pruned Collatz graph. Let us consider the $3n+1$ predecessor sets $\mathcal{P}_T(a)$ for $a \in \mathbb{N}$. We already know something about the structure of the full subgraph $P(a) := \langle \mathcal{P}_T(a) \rangle_T$ of the Collatz graph Γ_T generated by $\mathcal{P}_T(a)$:

(1) $P(a)$ is a tree, if and only if a is non-cyclic (remark 1.10).
(2) $P(a)$ is just an infinite line with terminal vertex a, if and only if $a \equiv 0$ mod 3 (corollary 3.5).

The first fact means that we can safely restrict our attention to trees, as non-cyclic numbers abound (we could also make a tree out of any $\mathcal{P}_T(a)$ by deleting the edge with initial vertex a, if there is one). Fact (2) implies that there are two main structures of $\mathcal{P}_T(a)$, depending on whether or not a is divisible by 3. This led Lagarias and Weiss [**LaW**, p. 241] to the idea of "pruning" the "$3x+1$ tree" by cutting away the vertices which are divisible by 3.

More precisely: take the $3n+1$ predecessors of 1, $\mathcal{P}_T(1) \subset \mathbb{N}$, form the full subgraph of the Collatz graph Γ_T generated by them (which is conjectured to be the whole of Γ_T), and remove the edge $(1,2)$ to obtain the $3x+1$ *tree* $\mathcal{T} = (V, E)$ with vertices $V := \mathcal{P}_T(1)$ and edges $E := \{(n, T(n)) : n \in V, n \neq 1\}$. Now the pruned $3x+1$ tree is defined by $\mathcal{T}^* = (V^*, E^*)$ with vertices

$$V^* := \{n \in \mathcal{P}_T(1) : n \not\equiv 0 \mod 3\}$$

and edges

$$E^* := \{(n, T(n)) : n \in V^*, n \neq 1\}.$$

In our setting, the following definition is appropriate:

5.4. DEFINITION. The *pruned Collatz graph* $\Gamma_T^* = (V_T^*, E_T^*)$ is constructed by removing all vertices of the Collatz graph Γ_T which are divisible by 3, and their adjacent edges.

Thus, $V_T^* = \mathbb{N} \setminus 3\mathbb{N}$. As $n \not\equiv 0 \mod 3$ clearly implies $T(n) \not\equiv 0 \mod 3$, the set of edges is just $E_T^* = \{(n, T(n)) : n \in V_T^*\}$.

5.5. DEFINITION. For $a \in V_T^* = \mathbb{N} \setminus 3\mathbb{N}$, the *pruned predecessor set* is given by

$$\mathcal{P}_T^*(a) := \{n \in \mathcal{P}_T(a) : n \not\equiv 0 \mod 3\} = \mathcal{P}_T(a) \cap V_T^* .$$

Pruned counting functions. Our next aim is to construct appropriate counting functions which are related to pruned predecessor sets $\mathcal{P}_T^*(a)$ in just the same way as the counting functions $e_\ell(k, a)$ defined in 4.1 are related to the "unpruned" predecessor sets $\mathcal{P}_T(a)$. We follow the method given in the first part of section 4.

5.6. DEFINITION. Let $a \in \mathbb{N}$. The *set of admissible vectors w.r.t. a for the pruned Collatz graph* is

$$\mathcal{E}^*(a) := \{s \in \mathcal{E}(a) : v_s(a) \not\equiv 0 \mod 3\} .$$

For $\ell, k \in \mathbb{N}_0$, we use again the notation (cf. 4.1)

$$\mathcal{E}_{\ell,k}^*(a) := \{s \in \mathcal{E}^*(a) : \ell(s) = \ell, |s| = k\} = \mathcal{E}_{\ell,k}(a) \cap \mathcal{E}^*(a) .$$

For fixed $\ell \in \mathbb{N}_0$, the *pruned counting function* is

$$e_\ell^* : \mathbb{N}_0 \times \mathbb{N} \to \mathbb{N}_0, \qquad e_\ell^*(k, a) := \left| \mathcal{E}_{\ell,k}^*(a) \right| .$$

The basic properties of the $e_\ell^*(k, a)$ are similar to those of the $e_\ell(k.a)$, with some significant differences.

5.7. LEMMA. *For $a, b \in \mathbb{N}$ and $j, k, \ell \in \mathbb{N}_0$, we have*

(a) *If $a \equiv 0 \mod 3$, then $e_\ell^*(k, a) = 0$.*
(b) *If $a \not\equiv 0 \mod 3$, then $e_0^*(k, a) = 1$.*
(c) *If $k \geqslant j$, then $e_\ell(k, a) \geqslant e_\ell^*(k, a) \geqslant e_\ell^*(j, a)$.*
(d) *If $a \equiv b \mod 3^{\ell+1}$, then $e_\ell^*(k, a) = e_\ell^*(k, b)$.*

PROOF. (a) follows from corollary 3.5: If $a \equiv 0 \mod 3$, then $\mathcal{E}(a) = \{(k) : k \in \mathbb{N}_0\}$. Because $v_{(k)}(a) = 2^k a \equiv 0 \mod 3$, this implies

$$\mathcal{E}^*(a) = \{s \in \mathcal{E}(a) : v_s(a) \not\equiv 0 \mod 3\} = \varnothing .$$

(b) If $a \not\equiv 0 \mod 3$, we have $\mathcal{E}_{0,k}(a) = \{(k)\} = \mathcal{E}_{\ell,k}^*(a)$.
(c) follows from the definition of $\mathcal{E}^*(a)$, and from part (b) of lemma 4.2.
To prove (d), consider, for $x \in \mathbb{N}$ and $j \in \{1, 2\}$, the sets

$$\mathcal{E}_{\ell,k}^{*j}(x) := \{s \in \mathcal{E}_{\ell,k}(x) : v_s(x) \equiv j \mod 3\} ,$$
$$\mathcal{E}_{\ell+1,k+1}^{o1}(x) := \{s \in \mathcal{E}_{\ell+1,k+1}^o(x) : s_1 \geqslant 1\} .$$

We then have the bijections

$$\mathcal{E}_{\ell,k}^{*1}(x) \quad \longleftrightarrow \quad \mathcal{E}_{\ell+1,k+1}^{o1}(x), \qquad s \quad \longleftrightarrow \quad (0, 1) \cdot s,$$

and

$$\mathcal{E}_{\ell,k}^{*2}(x) \quad \longleftrightarrow \quad \mathcal{E}_{\ell+1,k}^{o}(x), \qquad s \quad \longleftrightarrow \quad (0,0) \cdot s.$$

If $a \equiv b \mod 3^{\ell+1}$, apply lemma 3.1 to show that $\mathcal{E}_{\ell+1,k+1}^{o1}(a) = \mathcal{E}_{\ell+1,k+1}^{o1}(b)$ and $\mathcal{E}_{\ell+1,k}^{o}(a) = \mathcal{E}_{\ell+1,k}^{o}(b)$. Together with the above bijections, these prove the equations of cardinalities

$$\left| \mathcal{E}_{\ell,k}^{*1}(a) \right| = \left| \mathcal{E}_{\ell,k}^{*1}(b) \right| \qquad \text{and} \qquad \left| \mathcal{E}_{\ell,k}^{*2}(a) \right| = \left| \mathcal{E}_{\ell,k}^{*2}(b) \right|.$$

Now the disjoint set decomposition $\mathcal{E}_{\ell,k}^{*}(x) = \mathcal{E}_{\ell,k}^{*1}(x) \cup \mathcal{E}_{\ell,k}^{*2}(x)$, applied for $x = a$ and $x = b$, completes the proof. □

The basic relations between the number of predecessors in the pruned Collatz graph and the pruned counting functions is given in the following result:

5.8. THEOREM. *Let $a \in \mathbb{N}$ be non-cyclic in the pruned Collatz graph Γ_T^*, and put $\lambda = \log_2 3 - 1$. Then, for each real $x > 0$,*

$$Z_{\mathcal{P}_T^*(a)}(x) \geqslant \sum_{\ell=0}^{\infty} \sum_{k=0}^{\lfloor \log_2(x/a) + \lambda \ell \rfloor} e_\ell^*(k, a).$$

PROOF. An inspection of the proofs of lemma 4.6, theorem 4.7, an corollary 4.9 shows that these results remain valid if the Collatz graph Γ_T is replaced by the pruned Collatz graph Γ_T^*. The theorem in question is just the "pruned" analogue of corollary 4.9. □

Inductive construction of the pruned counting functions. According to lemma 4.3, the ordinary counting functions e_ℓ can be generated by induction on ℓ. The construction is also valid for the pruned counting functions e_ℓ^*, the only difference being that the initial values $e_0^*(k, a)$ and $e_0(k, a)$ do not coincide for $a \equiv 0 \mod 3$.

5.9. LEMMA. *Let $a \in \mathbb{N}$ and $k, \ell \in \mathbb{N}_0$, and put $e_\ell^*(j, q) := 0$ for $j \in \mathbb{N}_0$ and $q \in (\frac{1}{3}\mathbb{Z}) \setminus \mathbb{Z}$. Then*

$$e_{\ell+1}^*(k, a) = \sum_{j=0}^{k} e_\ell^* \left(k - j, \frac{2^{j+1}a - 1}{3} \right).$$

PROOF. The proof is almost the same as that of lemma 4.3. First extend the notation of definition 5.5 by putting $\mathcal{E}_{\ell,j}^*(q) := \varnothing$ for $j \in \mathbb{N}_0$ and $q \in (\frac{1}{3}\mathbb{Z}) \setminus \mathbb{Z}$. Then there is a one-one-correspondence

$$\mathcal{E}_{\ell+1,k}^*(a) \quad \longleftrightarrow \quad \bigcup_{j=0}^{k} \mathcal{E}_{\ell,k-j}^* \left(\frac{2^{j+1}a - 1}{3} \right),$$

where the union on the right hand side is a disjoint one. Here a vector

$$(s_0, \ldots, s_\ell, j) \in \mathcal{E}^*_{\ell+1,k}(a)$$

corresponds in a unique way to a vector

$$(s_0, \ldots, s_\ell) \in \mathcal{E}^*_{\ell,k-j}\left(v_{(0,j)}(a)\right) = \mathcal{E}^*_{\ell,k-j}\left(\frac{2^{j+1}a - 1}{3}\right),$$

and vice versa. The lemma now follows by definition 5.5. □

As in the case of the $e_\ell(k,a)$ (see corollary 4.4), it is possible to construct the values of $e^*_\ell(k,a)$ inductively, without reference to the pruned Collatz graph.

5.10. COROLLARY. *The sequence of pruned counting functions* $(e^*_\ell)_{\ell \in \mathbb{N}_0}$ *is uniquely determined by*

(i)
$$e^*_0(k,a) = \begin{cases} 0 & \text{if } a \equiv 0 \mod 3 \quad or \quad a \in \bigcup_{\nu=1}^\infty \frac{1}{3^\nu}\mathbb{Z}, \\ 1 & \text{if } a \equiv 1 \mod 3 \quad or \quad a \equiv 2 \mod 3. \end{cases}$$

(ii)
$$e^*_{\ell+1}(k,a) = \sum_{j=0}^{k} e^*_\ell\left(k-j, \frac{2^{j+1}a-1}{3}\right) \qquad \text{for } \ell \in \mathbb{N}_0.$$

Odd predecessors in the pruned Collatz graph. There are situations where it is appropriate to consider odd predecessors in the pruned Collatz graph.

5.11. DEFINITION. *Let* $a \in \mathbb{N}$. *The set of basic admissible vectors w.r.t.* a *of the pruned Collatz graph is*

$$\mathcal{E}^{o*}(a) := \mathcal{E}^o(a) \cap \mathcal{E}^*(a) = \{s \in \mathcal{E}(a) : v_s(a) \equiv 1,5 \mod 6\}.$$

Again, for $\ell, k \in \mathbb{N}_0$ we put

$$\mathcal{E}^{o*}_{\ell,k}(a) := \mathcal{E}_{\ell,k}(a) \cap \mathcal{E}^{o*}(a).$$

For fixed $\ell \in \mathbb{N}_0$, the *odd & pruned counting function* is

$$e^{o*}_\ell : \mathbb{N}_0 \times \mathbb{N} \to \mathbb{N}_0, \qquad e^{o*}_\ell(k,a) := \left|\mathcal{E}^{o*}_{\ell,k}(a)\right|.$$

5.12. LEMMA. *Let* $a \in \mathbb{N}$ *and* $\ell, k \in \mathbb{N}_0$.
(a) *The map*

$$\mathcal{E}^*_{\ell,k}(a) \longrightarrow \bigcup_{j=0}^{k} \mathcal{E}^{o*}_{\ell,j}(a), \qquad (s_0, s_1, \ldots, s_\ell) \mapsto (0, s_1, \ldots, s_\ell),$$

is a bijection.

(b) $\quad e^*_\ell(k,a) = \sum_{j=0}^{k} e^{o*}_\ell(j,a).$

PROOF. Imitate the proof of lemma 5.2. □

The next lemma generalizes lemma 4.3.

5.13. LEMMA. *Let $a \in \mathbb{N}$ and $j, \ell, x \in \mathbb{N}_0$ with $j < \ell$. Then*

$$e_\ell(x, a) = \sum_{k=0}^{x} \sum_{s \in \mathcal{E}_{j,k}^{o\bullet}(a)} e_{\ell-j}(x - k, v_s(a))$$

PROOF. Recall that $e_\ell(x, a) = |\mathcal{E}_{\ell,x}(a)|$, and let $t = (t_0, \ldots, t_\ell) \in \mathcal{E}_{\ell,x}(a)$. By our concatenation of integer vectors (definition 2.3), we can write

$$(t_0, \ldots, t_\ell) = (t_0, \ldots, t_{\ell-j}) \cdot (0, t_{\ell-j+1}, \ldots, t_\ell) \,.$$

Let $s := (0, t_{\ell-j+1}, \ldots, t_\ell)$ denote the last vector, and put $k := |s|$ (which implies $0 \leqslant k \leqslant x$). By definition 2.15 and corollary 2.10, we see that

$$s \in \mathcal{E}_{j,k}^{o}(a) \qquad \text{and} \qquad (t_0, \ldots, t_{\ell-j}) \in \mathcal{E}_{\ell-j, x-k}(v_s(a)) \,.$$

On the other hand, any integer $k \in \{0, \ldots, x\}$ and any pair of feasible vectors (r, s) with $s \in \mathcal{E}_{j,k}^{o}(a)$ and $r \in \mathcal{E}_{\ell-j, x-k}(v_s(a))$ give rise to an admissible vector $t := r \cdot s \in \mathcal{E}_{\ell,x}(a)$. This establishes the following one-one-correspondence:

$$(5.2) \qquad \mathcal{E}_{\ell,x}(a) \quad \longleftrightarrow \quad \bigcup_{k=0}^{x} \bigcup_{s \in \mathcal{E}_{j,k}^{o}(a)} \mathcal{E}_{\ell-j, x-k}(v_s(a)) \,,$$

where the union on the right hand side is clearly disjoint.

In that last union, we can replace $\mathcal{E}_{j,k}^{o}(a)$ by $\mathcal{E}_{j,k}^{o*}(a)$, for the following reason: Let $s \in \mathcal{E}_{j,k}^{o}(a) \setminus \mathcal{E}_{j,k}^{*}(a)$. Then, by definition 5.6, $v_s(a) \equiv 0 \mod 3$. By corollary 3.5 we conclude that $\mathcal{E}_{\ell-j, x-k}(v_s(a)) = \varnothing$, as we assumed that $\ell - j \geqslant 1$.

This completes the proof of the summation formula. $\square$

The following lemma does not mention odd predecessors of the pruned Collatz graph. It is recorded here because its statement and the essential idea of its proof is similar to that of the preceding lemma.

5.14. LEMMA. *Let $a \in \mathbb{N}$ and $j, k, \ell, x \in \mathbb{N}_0$ with $j < \ell$ and $k \leqslant x$. Then*

$$e_\ell(x, a) \geqslant \sum_{s \in \mathcal{E}_{j,k}^{*}(a)} e_{\ell-j}(x - k, v_s(a))$$

PROOF. Similar to the bijection of (5.2), we have an injective map

$$\bigcup_{s \in \mathcal{E}_{j,k}^{*}(a)} \mathcal{E}_{\ell-j, x-k}(v_s(a)) \quad \longrightarrow \quad \mathcal{E}_{\ell,x}(a)$$

mapping a pair r, s of vectors with $s \in \mathcal{E}_{j,k}^{*}(a)$ and $r \in \mathcal{E}_{\ell-j, x-k}(v_s(a))$ onto their concatenation $r \cdot s \in \mathcal{E}_{\ell,x}(a)$. Injectivity of this map follows from the uniqueness of terminal parts, cf. lemma 2.5: Each vector $s \in \mathcal{E}_{j,k}^{*}(a)$ fulfills $\|s\| = k + j$, and each vector $t \in \mathcal{E}_{\ell,x}(a)$ admits a unique decomposition with a terminal part with norm $k + j$. $\square$

6. Comparison with other approaches

The plan of this section is to give a precise comparison of our notation with that of those articles in the literature which pertain to the "tree-search method". The terminus "tree-search method" is used by Applegate and Lagarias [AL1] to refer to a method for deriving density estimates for $3n + 1$ predecessor sets by counting vertices of the Collatz graph. The tree-search approach has been initiated by Crandall [Cra] in 1978, and it is also considered by Sander [San] and in the above-mentioned article of Applegate and Lagarias. Concerning the derivation of density estimates for predecessor sets, the tree-search method has been superseded by far by the method of "Krasikov inequalities" (cf. section I.7). On the other hand, tree-search is like a microscope admitting a closer look at more subtle structures of the Collatz graph.

The plan of this section is to clarify the interconnections between the notation introduced here and that used in the above-mentioned articles. We give a presentation of the density estimates of Crandall and Sander, which are closely related. In addition, we give a brief discussion of the tree-search estimates of Applegate and Lagarias. We will see that all these estimates are built upon an estimation of

$$(6.1) \qquad \min_{a \in \mathbb{N}, 3 \nmid a} e_\ell^*(k, a) \qquad \text{for sufficiently many indices } k, \ell.$$

To clarify the implication of this fact, we introduce an appropriate notion of a "uniform bound", and we conclude that what has been proved by tree-search are *uniform lower bounds*.

Uniform bounds. The density estimates for $3n + 1$ predecessor sets which will be given here are derived through estimating series like that of theorem 5.3:

$$Z_{\mathcal{P}_T^\circ(a)}(x) \geqslant \sum_{\ell=0}^{\infty} e_\ell \left(\left\lfloor \log_2 \frac{x}{a} + \lambda \ell \right\rfloor, a \right).$$

Here the terms on the right hand side depend on x only through the quotient x/a. This motivates the following definition, which we give in an abstract setting.

6.1. DEFINITION. Let $U \subset \mathbb{N}$, and consider a family $(S(a))_{a \in U}$ of subsets $S(a) \subset \mathbb{N}$. A function $\varphi :]0, \infty[\to \mathbb{R}$ is called a *uniform lower bound* for the family $(S(a))_{a \in U}$, if there is real constant $\xi_0 > 0$ such that

$$Z_{S(a)}(x) \geqslant \varphi \left(\frac{x}{a} \right) \qquad \text{whenever} \qquad \frac{x}{a} \geqslant \xi_0 .$$

Similarly, φ is called a *uniform upper bound* for the family $(S(a))_{a \in U}$, if there is real constant $\xi_0 > 0$ such that

$$Z_{S(a)}(x) \leqslant \varphi \left(\frac{x}{a} \right) \qquad \text{whenever} \qquad \frac{x}{a} \geqslant \xi_0 .$$

6.2. REMARK. The conditions imposed on $\varphi :]0, \infty[\to \mathbb{R}$ in this definition can be rewritten in the following way: φ is a uniform lower bound for a family $(S(a))_{a \in U}$, if and only if

$$\liminf_{x \to \infty} \left(\inf_{a \in U} \frac{Z_{S(a)}(ax)}{\varphi(x)} \right) \geq 1 \, ;$$

this is easily seen by replacing x/a by x in the definition. Similarly, φ is a uniform upper bound for $(S(a))_{a \in U}$, if and only if

$$\limsup_{x \to \infty} \left(\sup_{a \in U} \frac{Z_{S(a)}(ax)}{\varphi(x)} \right) \leq 1 \, .$$

Here we are especially interested in uniform lower bounds of the type $\varphi(\xi) = \xi^c$, where c is some constant satisfying $0 < c < 1$, for the family of odd predecessor sets

$$(\mathcal{P}_T^o(a))_{a \in U} \qquad \text{where} \quad U = \mathbb{N} \setminus 3\mathbb{N} \, .$$

Uniform lower bounds for the smaller family $(\mathcal{P}_T^o(a))_{a \in V}$, where $V \subset U$ is the subset of non-cyclic numbers, will be derived by estimating the minimum (6.1). The following lemma shows that, in order to obtain a uniform lower bound of the type $\varphi_0(\xi) = \xi^c$ for the family indexed by U, it suffices to have a slightly larger uniform lower bound $\varphi_1(\xi) = \xi^{c+\varepsilon}$ for the smaller family indexed by the subset V of non-cyclic numbers.

6.3. LEMMA. *Let P denote one of the three 'predecessor set operators' $\mathcal{P}_T$, $\mathcal{P}_T^o$, $\mathcal{P}_T^*$, let $c, \varepsilon \in \mathbb{R}$ satisfy $0 < c < 1$ and $\varepsilon > 0$, and let $\varphi_1(\xi) = \xi^{c+\varepsilon}$ be a uniform lower bound for the family*

$$(P(a))_{a \in V} \qquad \text{where} \quad V = \{a \in \mathbb{N} : a \not\equiv 0 \bmod 3, \ a \text{ non-cyclic}\} \, .$$

Then $\varphi_0(\xi) = \xi^c$ is a uniform lower bound for the larger family $(P(a))_{a \in \mathbb{N} \setminus 3\mathbb{N}}$.

PROOF. Put $U := \mathbb{N} \setminus 3\mathbb{N}$. For any cyclic number $a \in U$, we know by lemma 4.10 that there is a non-cyclic number $\tilde{a} \in U$ satisfying both

$$a \leq \tilde{a} \leq 16\,a \qquad \text{and} \qquad \mathcal{P}_T(\tilde{a}) \subset \mathcal{P}_T(a) \, .$$

For any $b \in \mathbb{N}$, we have $\mathcal{P}_T^o(b) = \mathcal{P}_T(b) \cap (\mathbb{N} \setminus 2\mathbb{N})$ and $\mathcal{P}_T^*(b) = \mathcal{P}_T(b) \cap (\mathbb{N} \setminus 3\mathbb{N})$. This proves that $\tilde{a}$ fulfilles

$$a \leq \tilde{a} \leq 16\,a \qquad \text{and} \qquad P(\tilde{a}) \subset P(a) \, ,$$

for each predecessor set operator $P \in \{\mathcal{P}_T, \mathcal{P}_T^o, \mathcal{P}_T^*\}$. Taking into account the assumption that $\varphi_1(\xi) = \xi^{c+\varepsilon}$ is a uniform lower bound for the smaller family indexed by the non-cyclic numbers, we infer that, for any *cyclic* number $a \in U$ and any $x \in \mathbb{N}$,

$$Z_{P(a)}(x) \geq Z_{P(\tilde{a})}(x) \geq \left(\frac{x}{\tilde{a}} \right)^{c+\varepsilon} \qquad \text{whenever} \quad \frac{x}{\tilde{a}} \geq \xi_0 \, .$$

This implies

$$Z_{P(a)}(x) \geq \left(\frac{x}{a} \right)^c \qquad \text{whenever} \quad \frac{x}{a} \geq \max \left\{ 16\,\xi_0, \left(\frac{1}{16} \right)^{(c+\varepsilon)/\varepsilon} \right\},$$

which completes the proof. $\square$

Crandall's approach.   R. E. Crandall studied in [**Cra**] (1978) the $3n + 1$ problem via the map

$$C(m) := \frac{3m + 1}{2^{e(m)}} \qquad \text{for any positive odd integer } m,$$

where $e(m)$ is uniquely determined by the condition that $C(m)$ again should be an odd integer.

6.4. DEFINITION. Let m denote a positive odd integer. The *Crandall trajectory* of m is defined by

$$T^C(m) := \{C(m), C^2(m), \ldots\} = \{C^k(m) : k \in \mathbb{N}\} .$$

The *Crandall height* of m is defined by the cardinality

$$h(m) := \left| T_m^C \right| \quad \in \quad \{1, 2, 3, \ldots, \infty\} .$$

Clearly, $h(m)$ is finite if and only if the $3n + 1$ trajectory $T_f(m)$ eventually runs into a cycle. As 1 is fixed under C, we infer that $h(m)$ is finite if $1 \in \{C^j(m) : j = 1, 2, 3, \ldots\}$. Some examples: $h(1) = 1$, $h(7) = 5$, $h(2^{12} - 1) = 19794$ (data from [**Cra**]). For an odd positive integer m, the sequence $(C(m), C^2(m), \ldots)$ is precisely the sequence of odd integers in the $3n + 1$ trajectory $T_T(m)$ as given in definition 1.8. This implies that the $3n + 1$ conjecture is equivalent to

6.5. CONJECTURE. *Every positive odd integer has finite height.*

We now discuss that part of Crandall's machinery that leads to a density estimate for the odd predecessor set $\mathcal{P}_T^o(1) = \{m \in \mathbb{N} : m \text{ odd}, 1 \in T_m^C\}$.

6.6. DEFINITION. Define, for $a \in \mathbb{N}$ and $n \in \mathbb{Q}$, the function

$$B_a(n) := \frac{2^a n - 1}{3} .$$

Further define, for $k \in \mathbb{N}$ and $a_k, \ldots, a_1 \in \mathbb{N}$, inductively the functions

$$B_{a_k \ldots a_1}(n) := B_{a_k}(B_{a_{k-1} \ldots a_1}(n)) = \frac{2^a B_{a_{k-1} \ldots a_1}(n) - 1}{3} .$$

The relation between this notation and our notation based on admissible vectors is given by the following

6.7. LEMMA. *Let $a \in \mathbb{N}$ and $q \in \mathbb{Q}$. Then $B_a(q) = v_{(0,a-1)}(q)$. Moreover, if $k \in \mathbb{N}$ and $a_k, \ldots, a_1 \in \mathbb{N}$, then*

$$B_{a_k \ldots a_1}(q) = v_{(0, a_k - 1, \ldots, a_1 - 1)}(q) .$$

PROOF. By definition 2.9, we have

$$(6.2) \qquad v_{(0,a-1)}(q) = v_- \left(2^{a-1} q\right) \frac{2^a q - 1}{3} = B_a(q) .$$

To make the induction step, suppose $B_{a_{k-1} \ldots a_1}(q) = v_{(0, a_{k-1} - 1, \ldots, a_1 - 1)}(q)$. Then, by corollary 2.10 and (6.2),

$$v_{(0, a_k - 1, \ldots, a_1 - 1)}(q) = v_{(0, a_k - 1)}\left(v_{(0, a_{k-1} - 1, \ldots, a_1 - 1)}(q)\right)$$
$$= B_{a_k}\left(B_{a_{k-1} \ldots a_1}(q)\right) = B_{a_k \ldots a_1}(q) . \qquad \square$$

6.8. NOTATION. Crandall [**Cra**] introduces the following notation:

$$G := \left\{ (a_j, \dots, a_1) \,\middle|\, \begin{array}{l} j \in \mathbb{N}, \quad a_j, \dots, a_1 \in \mathbb{N}, \\ a_1 > 2, \quad 2^{a_1} \equiv 4,7 \pmod 9, \\ 2^{a_i} B_{a_{i-1} \dots a_1}(1) \equiv 4,7 \pmod 9 \quad \text{for } 2 \leqslant i \leqslant j-1 \\ 2^{a_j} B_{a_{j-1} \dots a_1}(1) \equiv 1 \pmod 3. \end{array} \right\}.$$

It is worth to single out the subset

$$G^* := \left\{ (a_j, \dots, a_1) \,\middle|\, \begin{array}{l} j \in \mathbb{N}, \quad a_j, \dots, a_1 \in \mathbb{N}, \\ a_1 > 2, \quad 2^{a_1} \equiv 4,7 \pmod 9, \\ 2^{a_i} B_{a_{i-1} \dots a_1}(1) \equiv 4,7 \pmod 9 \quad \text{for } 2 \leqslant i \leqslant j. \end{array} \right\}$$

For $j \in \mathbb{N}$ and $z \in \mathbb{R}$, $z > 0$, we also denote

$$G_{j,z} := \{ (a_j, \dots, a_1) \in G : a_1 + \cdots + a_j \leqslant z \} \quad \text{and} \quad G^*_{j,z} := G_{j,z} \cap G^*.$$

We shall see below that G is related to a certain set of admissible vectors. Note that the definition of G^* (which is not contained in Crandall's article [**Cra**]) can be restated in the following way:

$$G^* = \{ (a_j, \dots, a_1) \in G : B_{a_j \dots a_1}(1) \not\equiv 0 \pmod 3 \}.$$

For this reason, we expect that G^* is related to the pruned Collatz graph in just the same way as G is related to the complete Collatz graph.

6.9. LEMMA. *The following map*

$$\tau : G \to \mathcal{E}^o(4), \qquad \tau(a_j, \dots, a_1) = (0, a_j - 1, \dots, a_2 - 1, a_1 - 3)$$

is a bijection satisfying $B_{a_j \dots a_1}(1) = v_{\tau(a_j, \dots, a_1)}(4)$. *Moreover,* $\tau(G^*) = \mathcal{E}^{o*}(4)$ *and, for fixed* $j \in \mathbb{N}$ *and* $z \in \mathbb{R}$ *with* $z \geqslant j + 2$, *it has the images*

$$\tau(G_{j,z}) = \bigcup_{k=0}^{\lfloor z \rfloor - j - 2} \mathcal{E}^o_{j,k}(4) \quad \text{and} \quad \tau(G^*_{j,z}) = \bigcup_{k=0}^{\lfloor z \rfloor - j - 2} \mathcal{E}^{o*}_{j,k}(4).$$

PROOF. Let $(a_j, \dots, a_1) \in G$. From the definition of G, we have $a_1 > 2$, whence $(a_j, \dots, a_2, a_1 - 2) \in \mathbb{N}^j$. Using lemma 6.7,

$$B_{a_j \dots a_1}(1) = B_{a_j \dots a_2(a_1-2)}(4) = v_{(0, a_j-1, \dots, a_2-1, a_1-3)}(4) = v_{\tau(a_j, \dots, a_1)}(4).$$

Together with the definition of G, we infer the equivalences

$$
\begin{aligned}
(a_j, \dots, a_1) \in G \quad &\Longleftrightarrow \quad B_{a_j \dots a_1}(1) \in \mathbb{Z} \\
&\Longleftrightarrow \quad (0, a_j - 1, \dots, a_2 - 1, a_1 - 3) \in \mathcal{E}(4).
\end{aligned}
$$

This proves that τ is bijection. The equation $\tau(G^*) = \mathcal{E}^{o*}(4)$ is true because of the equivalences (valid for $(a_j, \ldots, a_1) \in G$)

$$(a_j, \ldots, a_1) \in G^* \quad \Longleftrightarrow \quad v_{\tau(a_j, \ldots, a_1)}(4) = B_{a_j \ldots a_1}(1) \not\equiv 0 \mod 3$$
$$\Longleftrightarrow \quad \tau(a_j, \ldots, a_1) \in \mathcal{E}^{o*}(4) \,.$$

To prove the two remaining equations, observe that

$$|\tau(a_j, \ldots, a_1)| = a_1 - 3 + \sum_{i=2}^{j}(a_i - 1) = a_1 + \ldots + a_j - j - 2 \,.$$

Hence $a_j + \ldots + a_1 \leqslant z \Leftrightarrow a_1 + \ldots + a_j - j - 2 \leqslant \lfloor z \rfloor - j - 2$, which completes the proof. $\square$

6.10. COROLLARY. *For $j \in \mathbb{N}$ and $z \in \mathbb{R}$ with $z \geqslant j + 2$, we have*

$$|G_{j,z}| = e_j(\lfloor z \rfloor - j - 2, 4) \quad \text{and} \quad |G_{j,z}^*| = e_j^*(\lfloor z \rfloor - j - 2, 4) \,.$$

PROOF. Combine the preceding lemma with lemma 5.2 or lemma 5.12, respectively. $\square$

Crandall's estimate. The first density estimate for $3n + 1$ predecessor sets is an inequality involving Crandall's counting function

$$\pi^C(x) := \left|\{m \in (2\mathbb{N} - 1) : m \leqslant x, h(m) < \infty\}\right| = Z_{\mathcal{P}_T^o(1)}(x) \,,$$

which counts the odd numbers $\leqslant x$ which satisfy the $3n + 1$ conjecture.

Theorem (6.1) of [Cra] asserts that there exists a positive constant c such that $\pi^C(x) > x^c$ for sufficiently large x. Based on Crandall's ideas, we shall prove here that the function $\varphi(\xi) = \xi^c$ for certain values of c is a uniform lower bound in the sense of definition 6.1 for the family $(\mathcal{P}_T^o(a))_{a \in \mathbb{N} \setminus 3\mathbb{N}}$.

Crandall's proof uses a series similar to our estimate given in corollary 4.9; this comes from the fact that he filtrates his counting function $\pi^C(x)$ according to heights.

6.11. DEFINITION. For a positive integer h, define the counting function

$$\pi_h^C(x) := \left|\{m \in (2\mathbb{N} - 1) : m \leqslant x, h(m) = h\}\right| \quad \text{for any } x \in \mathbb{R} \,.$$

The next lemma rewrites a step of Crandall's proof of his theorem (6.1).

6.12. LEMMA. *Let $x \in \mathbb{R}$, $x > 0$, let $h \in \mathbb{N}$, and put $\lambda = \log_2 3 - 1$. Then*

$$\pi_h^C(x) \geqslant e_h\left(\left\lfloor \log_2\left(\frac{x}{4}\right) + h\lambda \right\rfloor, 4\right) \,.$$

PROOF. (Compare the proof of theorem 5.3.) First let $(a_h, \ldots, a_1) \in G$ and put $s := \tau(a_h, \ldots, a_1)$. By lemma 6.9 and lemma 2.13 we know that

$$B_{a_h \ldots a_1}(1) = v_s(4) = c(s) \cdot 4 - r(s) \leqslant 4c(s) = \frac{4 \cdot 2^{h+|s|}}{3^h} = \frac{2^{a_1 + \ldots + a_h}}{3^h}.$$

If we put $z := \log_2(3^h x) = \log_2 x + h \log_2 3$, then this implies that $B_{a_h \ldots a_1}(1) \leqslant x$ for every sequence $(a_h, \ldots, a_1) \in G_{h,z}$. Together with the bijection of lemma 6.9, theorem 2.19 (which is a generalization of Crandall's "uniqueness theorem" (4.1) of [**Cra**]) yield the first estimate in

$$\pi_h^C(x) \geqslant |G_{h,z}| = e_h(\lfloor z \rfloor - h - 2, 4)$$
$$= e_h(\lfloor \log_2 x + h \log_2 3 \rfloor - h - 2, 4)$$
$$= e_h\left(\left\lfloor \log_2\left(\frac{x}{4}\right) + h\lambda \right\rfloor, 4\right);$$

the first equation is due to corollary 6.10, whereas the remaining equations are just computation. $\square$

As possible heights run from 1 to ∞, we have an estimate by a series:

$$Z_{\mathcal{P}_T^o(1)}(x) = \pi^C(x) = \sum_{h=1}^{\infty} \pi_h^C(x) \geqslant \sum_{h=1}^{\infty} e_h\left(\left\lfloor \log_2\left(\frac{x}{4}\right) + h\lambda \right\rfloor, 4\right).$$

This is also an immediate consequence of theorem 5.3, as $Z_{\mathcal{P}_T^o(1)}(x) \geqslant Z_{\mathcal{P}_T^o(4)}(x)$.

In his lemma (5.2), Crandall states a lower bound for the cardinalities of sets of the type $G_{j,z}$ defined in 6.8 above. An examination of his proof shows that, in fact, he produced a lower bound for the pruned counting functions.

6.13. LEMMA. *Let* $j \in \mathbb{N}$ *and* $z \in \mathbb{R}$ *such that* $z \geqslant 0$. *Then:*

$$\min_{a \in \mathbb{N}, 3 \nmid a} e_j^*(\lfloor z \rfloor - j, a) \geqslant \left(2 \cdot \left\lfloor \frac{z}{6j} \right\rfloor\right)^j.$$

PROOF. We make essential use of Crandall's idea, cf. the proof of lemma (5.2) in [**Cra**].

Let $a \in \mathbb{N}$ be such that $3 \nmid a$, and let

$$A_\nu(a) := \left\{ (0, s_\nu, \ldots, s_j) \in \bigcup_{k=0}^{\lfloor z \rfloor - j} \mathcal{E}_{j-\nu+1,k}^{o*}(a) : s_i \leqslant \frac{z}{j} - 1 \text{ for } i = \nu, \ldots, j \right\}.$$

Using part (b) of lemma 5.12, we have then

(6.3) $$e_j^*(\lfloor z \rfloor - j, a) = \sum_{k=0}^{\lfloor z \rfloor - j} e_j^{o*}(k, a) \geqslant A_1(a).$$

Now we proceed by backward induction on ν starting with $\nu = j$ and ending with $\nu = 1$, and we shall prove the claim

$$(6.4) \qquad |A_\nu(a)| \geq \left(2 \cdot \left\lfloor \frac{z}{6j} \right\rfloor\right)^{j-\nu+1} \qquad \text{for each } a \in \mathbb{N} \text{ with } 3 \nmid a.$$

Let $\nu := j$. In order to fulfill $(0, s_j) \in A_j(a)$, what is required for s_j are the three conditions

$$s_j \leq \frac{k}{j} \leq \frac{z}{j} - 1, \qquad v_{(0,s_j)}(a) \in \mathbb{Z}, \qquad \text{and} \qquad v_{(0,s_j)}(a) \not\equiv 0 \mod 3.$$

Via $v_{(0,s_j)}(a) = (2^{s_j+1}a-1)/3$, the latter two conditions amount to $2^{s_j+1}a \equiv 4,7$ mod 9. This congruence has precisely two solutions s_j out of six consecutive integers, hence the number of possible choices s_j is at least $2 \cdot \lfloor z/6j \rfloor$. This proves

$$|A_j(a)| \geq 2 \cdot \left\lfloor \frac{z}{6j} \right\rfloor,$$

which is (6.4) for $\nu = j$.

Now let ν denote an integer satisfying $1 \leq \nu \leq j - 1$. For each given vector $(0, s_{\nu+1}, \ldots, s_j) \in A_{\nu+1}(a)$, we want to construct vectors

$$(0, s_\nu, \ldots, s_j) \in A_\nu(a).$$

To abbreviate formulae, put $a_\nu := v_{(0,s_{\nu+1},\ldots,s_j)}(a)$. What is required for s_ν are precisely the conditions

$$(6.5) \qquad s_\nu \leq \frac{k}{j} \leq \frac{z}{j} - 1 \qquad \text{and} \qquad 2^{s_\nu+1} a_\nu \equiv 4,7 \mod 9.$$

As before, there are precisely two possible choices of s_ν out of six consecutive integers, and the arguments above together with the induction hypothesis (6.4) (for $\nu + 1$) imply

$$|A_\nu(a)| \geq 2 \cdot \left\lfloor \frac{z}{6j} \right\rfloor \cdot |A_{\nu+1}(a)| \geq \left(2 \cdot \left\lfloor \frac{z}{6j} \right\rfloor\right)^{j-\nu+1}.$$

The claim of the lemma now follows from (6.3) and estimate (6.4) for $\nu = 1$. $\quad\square$

Crandall's proof of his predecessor density estimate makes use of the following fact.

6.14. LEMMA.

$$\lim_{t \to \infty} e^{-t} \sum_{k=0}^{\lfloor t \rfloor} \frac{t^k}{k!} = \frac{1}{2}.$$

This result is well-known. For a discussion of it, and further references, see B. C. Berndt [Ber] (the discussion of "Entry 48" of Ramanujan's notebooks in chapter 12, p. 181f). A very intuitive proof is the following [**Ber**, p. 182]: Suppose that each of the n independent random variables X_k, $1 \leqslant k \leqslant n$, has a Poisson distribution with parameter 1. Then $S_n = \sum_{k=1}^{n} X_k$ has a Poisson distribution with parameter n. Thus,

$$P(S_n \leqslant n) = e^{-n} \sum_{k=0}^{n} \frac{n^k}{k!} \, .$$

After applying the central limit theorem, we conclude that

$$\lim_{n \to \infty} P(S_n \leqslant n) = \tfrac{1}{2} \, .$$

6.15. LEMMA. *Let* $a \in \mathbb{N} \setminus 3\mathbb{N}$ *be non-cyclic in the Collatz graph, and fix* $\lambda = \log_2 \frac{3}{2}$. *Suppose there are positive real constants* c_0, r *such that*

$$e_h^* \left(\left\lfloor \log_2 \frac{x}{a} + h\lambda \right\rfloor, a \right) \geqslant c_0 \frac{(r \log_2(x/a))^h}{h!} \, , \qquad if \quad 1 \leqslant h \leqslant r \log_2 \frac{x}{a} \, .$$

Then, for any positive constant $c < \dfrac{r}{\log 2}$, *there is a constant* $\xi_0 > 0$ *such that*

$$Z_{\mathcal{P}_T^o(a)}(x) \geqslant \left(\frac{x}{a} \right)^c \qquad for \quad \frac{x}{a} > \xi_0 \, .$$

PROOF. (Cf. [**Cra**], proof of theorem (5.1).) Theorem 5.3 and the assumptions combine to give

$$Z_{\mathcal{P}_T^o(a)}(x) \geqslant \sum_{h=1}^{\infty} e_h^* \left(\left\lfloor \log_2 \frac{x}{a} + h\lambda \right\rfloor, a \right) \geqslant c_0 \sum_{h=1}^{\lfloor r \log_2(x/a) \rfloor} \frac{(r \log_2(x/a))^h}{h!} \, .$$

Now the idea is to use the preceding lemma with t replaced by $r \log_2 x$. This proves: for any $\varepsilon > 0$ there is an $\xi_1(\varepsilon) > 0$ such that

$$Z_{\mathcal{P}_T^o(a)}(x) > \left(\tfrac{1}{2} - \varepsilon \right) c_0 \, e^{r \log_2(x/a)} = c_0 \left(\frac{1}{2} - \varepsilon \right) \left(\frac{x}{a} \right)^{r/\log 2} \qquad for \quad \frac{x}{a} \geqslant \xi_1(\varepsilon) \, .$$

Put $\varepsilon := \tfrac{1}{4}$ and choose $\xi_0 > \xi_1(\varepsilon)$ sufficiently large, and the claim follows. $\square$

6.16. LEMMA. *Let* $h \in \mathbb{N}$, $a \in \mathbb{N} \setminus 3\mathbb{N}$, *and put* $r := \left(3 + 2e - \log_2 3 \right)^{-1}$, *where* $e = \exp(1)$ *is Euler's number. Then*

$$e_h^* \left(\left\lfloor \log_2 \frac{x}{a} + h\lambda \right\rfloor, a \right) \geqslant \frac{(r \log_2(x/a))^h}{h!} \qquad for \quad \frac{x}{a} \geqslant 2^{h/r} \, .$$

PROOF. This is similar to theorem (5.1) of [**Cra**], we give a restructured version of Crandall's proof. We shall demonstrate the chain of estimates

$$(6.6) \quad e_h^*(\lfloor z \rfloor - h, a) \geqslant \left(\frac{\log_2(3^h x/a)}{3h} - 2 \right)^h \geqslant \left(\frac{er}{h} \log_2 \frac{x}{a} \right)^h \geqslant \frac{(r \log_2(x/a))^h}{h!} \, ,$$

where $z := \log_2(3^h x/a)$.

The last '$\geqslant$' is just an application of Stirling's estimate $h! > (h/e)^h$. The first one follows from lemma 6.13 in the following way:

$$e_h^*(\lfloor z \rfloor - h, a) \geqslant \left(2 \cdot \left\lfloor \frac{z}{6h} \right\rfloor \right)^h .$$

Observing $\lfloor \xi \rfloor \geqslant \xi - 1$ for any real ξ and inserting z, we find

$$e_h^*(\lfloor z \rfloor - h, a) \geqslant \left(\frac{\log_2(x/a) + h \log_2 3}{3h} - 2 \right)^h ,$$

which is the first '$\geqslant$' of (6.6). For the middle inequality, it remains to prove

$$\frac{\log_2(3^h x/a) - 6h}{3h} \geqslant \frac{er}{h} \log_2 \frac{x}{a} .$$

This amounts to

$$\log_2 \left(\left(\tfrac{3}{64}\right)^h \frac{x}{a} \right) \geqslant 3er \log_2 \frac{x}{a} \qquad \text{or} \qquad \log_2 \frac{x}{a} \geqslant \frac{h \log_2 \left(\tfrac{3}{64}\right)}{3er - 1} .$$

But this is equivalent to the condition $x/a \geqslant 2^{h/r}$, provided we choose r such that

$$r = \frac{3er - 1}{\log_2\left(\tfrac{3}{64}\right)} , \qquad \text{i.e.} \qquad r := \frac{1}{3(e+2) - \log_2 3} . \qquad \square$$

6.17. THEOREM. *For each positive constant*

$$c < \frac{r}{\log 2} = \frac{1}{3(e+2)\log 2 - \log 3} \doteq 0.1147 ,$$

the function $\varphi(\xi) = \xi^c$ is a uniform lower bound for the family $(P_T^o(a))_{a \in \mathbb{N} \setminus 3\mathbb{N}}$. In particular, $\pi^C(x) > x^c$ for sufficiently large x.

PROOF. Combine the lemmas $6.15, 6.16$, and 6.3. For the estimate concerning Crandall's counting function, observe that $\pi^C(x) = Z_{\mathcal{P}_T^o(1)}(x)$. $\square$

Note that the constant $c < r/\log 2 \approx 0.1147$ computed here is much larger than the value $c \approx 0.057$ computed by Sander [San] for Crandall's estimate. Here a factor 2 is gained just because it is more appropriate to consider the sets $\mathcal{P}_T^o(a)$ for $a \in \mathbb{N} \setminus 3\mathbb{N}$, instead of restricting attention to the predecessors of 1.

Sander's estimate. In 1987, J. W. Sander (see [San]) improved Crandall's estimate. We give essentially Sander's main lemma; the proof relies on a nice development starting with Crandall's central idea given in the proof of lemma 6.13.

6.18. LEMMA. *Let $h, k \in \mathbb{N}$ with $k \geqslant 3h$. Then*

$$e_h^*(k, a) \geqslant \binom{\lfloor k/3 \rfloor}{h} \qquad \text{for any} \quad a \in \mathbb{N} \setminus 3\mathbb{N}.$$

PROOF. (Cf. [San], proof of the "main lemma".) Fix $a \in \mathbb{N} \setminus 3\mathbb{N}$. We construct a bijection

$$\psi : \{s \in \mathcal{E}^{o*}(a) : \ell(s) = h\} \to (3\mathbb{N})^h, \qquad \psi(s) = (\psi_1(s), \ldots, \psi_h(s))$$

with the property

$$(6.7) \qquad\qquad |s| \leqslant \psi_1(s) + \ldots + \psi_h(s).$$

To construct ψ, let $s = (0, s_1, \ldots, s_h) \in \mathcal{E}^{o*}(a)$, and put $a_j := v_{(0, s_{j+1}, \ldots, s_h)}(a)$ for $1 \leqslant j \leqslant h - 1$, and $a_h := a$. As in the proof of lemma 6.13, second condition of (6.5), we know that

$$2^{s_j + 1} a_j \equiv 4, 7 \mod 9 \qquad \text{for} \quad j = 1, \ldots, h.$$

Thus: in each interval $J_k := \{6k, 6k + 1, \ldots, 6k + 5\}$, $k \in \mathbb{N}$, there are precisely two possible values of s_j, say $s_j'(k) < s_j''(k)$, with the property $s_j''(k) - s_j'(k) \geqslant 2$. This implies $s_j''(k) \leqslant 6k + 5$ and $s_j'(k) \leqslant 6k + 3$. Now put

$$\psi_j(s) := \begin{cases} 6k + 3 & \text{if } s_j = s_j'(k), \\ 6k + 5 & \text{if } s_j = s_j''(k). \end{cases}$$

Then $s_j \leqslant \psi_j(s)$ for each $j = 1, \ldots, h$, and it is clear that ψ is a bijection satisfying (6.7). Now

$$
\begin{aligned}
e_h^*(k, a) &= |\{s \in \mathcal{E}^{o*}(a) : \ell(s) = h, |s| \leqslant k\}| && \text{by lemma 5.12,} \\
&\geqslant |\{(\psi_1, \ldots, \psi_h) \in (3\mathbb{N})^h : \psi_1 + \ldots + \psi_h \leqslant k\}| && \text{by (6.7),} \\
&= \left|\left\{(k_1, \ldots, k_h) \in \mathbb{N}^h : k_1 + \ldots + k_h \leqslant \frac{k}{3}\right\}\right| \\
&= \binom{\lfloor k/3 \rfloor}{h},
\end{aligned}
$$

where the last equation is elementary combinatorics. $\square$

A computation using

$$h!\binom{\lfloor k/3 \rfloor}{h} = \left\lfloor \frac{k}{3} \right\rfloor \cdot \left(\left\lfloor \frac{k}{3} \right\rfloor - 1\right) \cdot \ldots \cdot \left(\left\lfloor \frac{k}{3} \right\rfloor - h + 1\right) \geqslant \left(\frac{k}{3} - h\right)^h$$

shows that we can choose $r = (6 - \lambda)^{-1} \approx \frac{2}{11}$ in lemma 6.15, which implies that we can choose

$$c < \frac{r}{\log 2} = \frac{1}{7 \log 2 - \log 3} \doteq 0.2664,$$

in theorem 6.17, which is Sander's result $c = \frac{1}{4}$ [San].

Minorant vectors of Applegate and Lagarias. The article of Applegate and Lagarias [**AL1**] (1995) pertaining to the tree-search method is different in flavour from the articles discussed above. In fact, they do not give lower bounds for $e_\ell^*(k, a)$ as written up in lemmas 6.13 and 6.18. They estimate instead sums of the type

$$\sum_{\ell \geqslant \mu j} e_\ell^*(j - \ell, a)$$

where $j \in \mathbb{N}$ and $\mu \in \mathbb{R}$ with $0 < \mu < \log 2 / \log 3$ are fixed. We present that approach to the extend needed for seeing that what is proved in [**AL1**] is, actually, a uniform lower bound for the family $(\mathcal{P}_T^*(a))_{a \in \mathbb{N} \setminus 3\mathbb{N}}$.

Applegate and Lagarias first consider the following subgraphs of the pruned Collatz graph. For $k, a \in \mathbb{N}$ let $\mathcal{T}_k^*(a)$ denote the full subgraph of Γ_T^* generated by the set of vertices

$$\mathcal{V}_k^*(a) := \left\{ x \in \mathbb{N} : T^j(x) = a \text{ for some } j \leqslant k \right\},$$

where T is the $3n + 1$ function. In our terminology (cf. lemma 2.17), we have

$$\mathcal{V}_k^*(a) = \{v_s(a) : s \in \mathcal{E}^*(a), \|s\| \leqslant k\}.$$

Note that $\mathcal{T}_k^*(a)$ is a tree, if and only if a is non-cyclic.

The next step in [**AL1**] is to associate a *weight* to each vertex of $\mathcal{T}_k^*(a)$ such that weight(x) counts the edges of the path from x to a which arise from the odd branch of the $3n + 1$ function. Hence, for a leaf $x = v_s(a)$ with $T^k(x) = a$, we have

$$\text{weight}(x) = \text{weight}(v_s(a)) = \ell(s).$$

The link to predecessor sets is derived through the sets

$$V_j^*(a, \mu) := \left\{ x \in \mathbb{N} : T^j(x) = a, \text{weight}(x) \geqslant \mu j \right\}$$
$$= \{v_s(a) : s \in \mathcal{E}^*(a), \|s\| \leqslant k\},$$

where μ will be a delicately chosen real number satisfying $0 < \mu < \log 2 / \log 3$. Let us suppose now that a is non-cyclic. Then the cardinality of such a set is

$$N_j^*(a, \mu) := \left| V_j^*(a, \mu) \right| = \sum_{\ell \geqslant \mu j} e_\ell^*(j - \ell, a).$$

(In [**AL1**], we find the letter α instead of μ; here we prefer μ as we have the reservation $\alpha = \log_2 3$.) For $x = v_s(a) \in V_j^*(a, \mu)$, we conclude from lemma 2.13 that

$$\frac{x}{a} = \frac{v_s(a)}{a} \leqslant c(s) = \frac{2^j}{3^{\ell(s)}} \leqslant 2^j \cdot 3^{\mu j} = \exp\left(j(\log 2 - \mu \log 3)\right),$$

which is relation (3.1) of [**AL1**]. For arbitrary predecessors x of a in the pruned Collatz graph, i.e. for $x \in \mathcal{P}_T^*(a)$, we put

$$j(x) := \left\lfloor \frac{1}{\log 2 - \mu \log 3} \cdot \log \frac{x}{a} \right\rfloor.$$

If $2 \cdot 3^{-\mu} > 1$ (which is just the restriction on μ mentioned above), this gives

$$\frac{y}{a} \leqslant \left(2 \cdot 3^{-\mu}\right)^{j(x)} \leqslant \frac{x}{a} \leqslant \left(2 \cdot 3^{-\mu}\right)^{j(x)+1} \qquad \text{for} \quad y \in V^*_{j(x)}(a, \mu),$$

which implies $Z_{\mathcal{P}^*_T(a)}(x) \geqslant N^*_{j(x)}(a, \mu)$. To have an estimate like that of lemma 6.15, define $\gamma(x)$ such that

$$N^*_{j(x)}(a, \mu) = \left(2 \cdot 3^{-\mu}\right)^{(j(x)+1)\gamma(x)} \geqslant \left(\frac{x}{a}\right)^{\gamma(x)}.$$

Clearly,

$$\gamma(x) := \frac{1}{\log 2 - \mu \log 3} \cdot \frac{1}{j(x) + 1} \, \log N^*_{j(x)}(a, \mu)$$

does the job; if $x \to \infty$, then $j(x) \to \infty$, and

$$(6.8) \qquad \gamma_0 := \frac{1}{\log 2 - \mu \log 3} \, \liminf_{j \to \infty} \frac{1}{j} \left(\inf_{a \in \mathbb{N} \backslash 3\mathbb{N}} \log N^*_j(a, \mu) \right)$$

has the property that, for any $\varepsilon > 0$, there is a $\xi_0(\varepsilon) > 0$ such that, for $x \in \mathbb{N}$ and non-cyclic $a \in \mathbb{N} \backslash 3\mathbb{N}$,

$$Z_{\mathcal{P}^*_T(a)}(x) \geqslant \left(\frac{x}{a}\right)^{\gamma_0 - \varepsilon} \qquad \text{whenever} \quad \frac{x}{a} \geqslant \xi_0(\varepsilon).$$

In other words: *any* real number $\gamma < \gamma_0$ gives rise to a uniform lower bound $\varphi(\xi) = \xi^\gamma$ (in the sense of definition 6.1) for the family of predecessor sets $\{\mathcal{P}^*_T(a) : a \in \mathbb{N} \backslash 3\mathbb{N} \text{ non-cyclic}\}$. Using lemma 6.3, we infer that, for any $\gamma < \gamma_0$, $\varphi(\xi) = \xi^\gamma$ is a uniform lower bound for the family $\{\mathcal{P}^*_T(a) : a \in \mathbb{N} \backslash 3\mathbb{N}\}$.

Aiming at the construction of a lower bound for γ_0, Applegate and Lagarias introduce in [**AL1**] the weight counts (for non-cyclic a)

$$w^k_j(a) := \left| \{ x \in \mathbb{N} \backslash 3\mathbb{N} : T^k(x) = a, \text{weight}(x) = j \} \right|$$
$$= \left| \{ v_s(a) : s \in \mathcal{E}^*(a), \|s\| = k, \ell(s) = j \} \right| = \left| \mathcal{E}^*_{j,k-j}(a) \right| = e^*_j(k - j, a),$$

leading to the *vector of weights*

$$\mathbf{w}^*_k(a) := \left(w^k_0(a), \ldots, w^k_k(a) \right) = \left(e^*_0(k, a), e^*_1(k - 1, a), \ldots, e^*_k(0, a) \right).$$

A vector $\mathbf{w} = (w_0, \ldots, w_k)$ is said to *minorize* another vector $\mathbf{w}' = (w'_0, \ldots, w'_k)$, if

$$\sum_{j=0}^i w_{k-j} \leqslant \sum_{j=0}^i w'_{k-j} \qquad \text{for} \quad 0 \leqslant i \leqslant k.$$

The *minorant vector*

$$\mathbf{w}^-(k) := \left(w^-_0(k), \ldots, w^-_k(k) \right)$$

is determined by the conditions

$$\sum_{j=0}^{i} w_{k-j}^{-}(k) = \min_{a \in \mathbb{N} \setminus 3\mathbb{N}} \left\{ \sum_{j=0}^{i} e_{k-j}^{*}(j, a) \right\} \qquad \text{for} \quad 0 \leqslant i \leqslant k.$$

Observe that we know by lemma 5.7 that $e_{k-j}^{*}(j, a)$ only depends on the residue class of a to modulus 3^{k-j+1}, hence it suffices to take the minimum over the finite set

$$A_{k+1}^{*} := \{0 < a < 3^{k+1}, \ a \not\equiv 0 \mod 3\}.$$

The next step explicitly makes use of the fact that the subtreee had been chosen in the *pruned* Collatz graph. Suppose that we have a minorant vector for some k, say $k = 30$, and suppose that we know all possible trees $T_k^{*}(a)$, where a runs through the finite set A_{k+1}^{*}. Then a tree $T_{2k}^{*}(a)$ emerges by attaching a tree $T_k^{*}(b)$ to each leaf b of $T_k^{*}(a)$. Moreover, it is possible to compute a vector minorizing the minorant vector for $T_{2k}^{*}(a)$ by taking the *convolution*

$$\mathbf{w}^{-}(k) * \mathbf{w}^{-}(k) = \left(x_0^k(2), \ldots, x_k^k(2)\right),$$

where

$$x_i^k(2) = \sum_{i_1 + i_2 = i} w_{i_1}^{-}(k) w_{i_2}^{-}(k).$$

Iterating this j times, it is possible to obtain a lower bound

(6.9)
$$\sum_{i > jk\mu} x_i^k(j) \leqslant N_{jk}^{*}(a, \mu),$$

which does not depend on a. The remaining part to give a lower bound γ_k for γ_0 in (6.8) is motivated by the investigation of a stochastic model, see [LaW]. The limits of the sum in (6.9) needed to estimate (6.8) are computed using Chernoff's "large deviation bounds".

The method is computer-intensive. Applegate and Lagarias constructed via computer all possible trees $T_k^{*}(a)$ for $k = 1, \ldots, 30$, to obtain the minorant vector for these values of k. The final result is $\gamma_{30} = 0.6547$, giving the uniform lower bound $\varphi(\xi) = \xi^{0.6547}$, i.e. there is a constant ξ_0 which only depends on the construction, but does not depend on a, such that

$$Z_{\mathcal{P}_k^{*}(a)}(x) \geqslant \left(\frac{x}{a}\right)^{0.6547} \qquad \text{for } x \in \mathbb{N}, \ a \in \mathbb{N} \setminus 3\mathbb{N} \text{ satisfying } \frac{x}{a} \geqslant \xi_0.$$

3-ADIC AVERAGES OF COUNTING FUNCTIONS

The starting point of this chapter is the observation made in lemma II.4.2 that the counting functions for admissible vectors

$$e_\ell(k, a) = |\mathcal{E}_{\ell,k}(a)|$$

depend on their second variable only through its residue class to modulus 3^ℓ. This implies that we are concerned with functions

$$e_\ell : \mathbb{N}_0 \times \mathbb{Z}_3 \to \mathbb{N}_0 \qquad \text{for each} \quad \ell \in \mathbb{N}_0.$$

Moreover, these functions can be constructed recursively without direct reference to the Collatz graph. The counting functions e_ℓ are connected to $3n + 1$ predecessor estimates via the estimating series

$$s_n : \mathbb{Z}_3^* \to \overline{\mathbb{N}}_0, \qquad s_n(a) := \sum_{\ell=1}^{\infty} e_\ell(n + \lfloor \lambda \ell \rfloor, a),$$

here $\lambda = \log_2(\frac{3}{2})$ is a constant occurring frequently. After giving the basics of 3-adic numbers in section 1, this estimating series is introduced in section 2. Theorem 2.5 links the asymptotic behaviour of $s_n(a)$ when n tends to ∞ to the asymptotic behaviour of the predecessor counting function $Z_a(x)$ for $x \to \infty$, provided a is non-cyclic. Theorem 2.7 proves that all functions s_n are discontinuous on $\mathbb{Z}_3^*$. Note that Applegate and Lagarias [AL3] also use this set of invertible 3-adic integers in a $3n + 1$ context, but in a different way.

Though the s_n are topologically ill-behaved as functions on the topological group $\mathbb{Z}_3^*$, we shall see in section 3 that they are perfectly integrable w.r.t. the normalized Haar measure, and that we can compute their integrals to

$$\bar{s}_n = \int_{\mathbb{Z}_3^*} s_n(a) \, da = \sum_{\ell=1}^{\infty} \frac{1}{2 \cdot 3^{\ell-1}} \binom{n + \lfloor \alpha \ell \rfloor}{\ell}$$

where $\alpha = \log_2 3 = \lambda + 1$. These series (for $n \in \mathbb{N}$) are then called *averaged estimating sums*.

In section 4 we locate the maximals terms of these sums, i.e. we compute in theorem 4.3 the index $\ell_{\max}(n)$ of the maximal term asymptotically to first order. It turns out that $\ell_{\max}(n) \sim \ell_0(n)$ where $\ell_0(n)$ satisfies the relation

$$n + \lfloor \alpha \ell_0(n) \rfloor = 2 \ell_0(n),$$

i.e. for $\ell = \ell_0(n)$, the upper entry in the binomial coefficient in the averged estimating sum is just twice the lower entry. Theorem 4.3 says, roughly, that for a 'randomly chosen' $3n + 1$ iteration $(n, T(n), \ldots, T^k(n))$, it is most likely that it contains approximately $k/2$ steps through the branch $T_0(x) = x/2$, where the remaining approximately $k/2$ steps are effected by the branch $T_1(x) = (3x+1)/2$. This supports the usual $3n + 1$ heuristics, cf. [**Wag**].

In the final section 5, we prove a theorem about the asymptotic behaviour, for $n \to \infty$, of the averaged estimating series. This proof uses a close analysis of the asymptotic behaviour of binomial coefficients. Theorem 5.2 leads to the following conclusion: *If, for large n, the n-th estimating series $s_n(a)$, evaluated at a non-cyclic number $a \in \mathbb{N}$, is very close to the averaged n-th estimating series $\bar{s}_n$, then the $3n + 1$ predecessor set $\mathcal{P}_T(a)$ has positive asymptotic density.* The chapter ends with a heuristic remark why the $3n + 1$ problem should be special among '$pn + 1$' problems for larger primes p.

1. Basics of 3-adic numbers

In this section, we give the basic definitions and facts on 3-adic numbers which are needed in the subsequent sections, and in chapter IV. The topics discussed include 3-adic metric, representation of 3-adic numbers as sequences of digits, embedding of the integers into the 3-adic integers, residue classes, addition, multiplication, the set $\mathbb{Z}_3^*$ of invertible 3-adic integers, and Haar measures. The material is presented without proofs; for proofs and more background information we refer to standard literature on g-adic numbers, e.g. Hensel [**Hen**].

1.1. DEFINITION. For any $n \in \mathbb{N}$, denote by $v_3(n)$ the exponent of 3 in the decomposition of n into prime factors. For a rational number $x \neq 0$, say $x = \pm r/s$ with $r, s \in \mathbb{N}$, put $v_3(x) := v_3(r) - v_3(s)$. Finally, put $v_3(0) := \infty$. The 3-*adic valuation* on $\mathbb{Q}$ is defined by $|x|_3 := 3^{-v_3(x)}$ (with $3^{-\infty} = 0$).

1.2. LEMMA & DEFINITION. *The 3-adic metric on the rationals is defined by*

$$d_3 : \mathbb{Q} \times \mathbb{Q} \longrightarrow \mathbb{R}_+, \qquad d_3(x, y) := |x - y|_3 \, .$$

It is a non-Archimedean metric, i.e. $d_3(x, y) = 0 \Leftrightarrow x = y$, $d_3(x, y) = d_3(y, x)$, and $d_3(x, z) \leqslant \max\{d_3(x, y), d_3(y, z)\}$ for any $x, y, z \in \mathbb{Q}$. The set $\mathbb{Q}_3$ of 3-adic reals is defined to be the completion of $\mathbb{Q}$ by the 3-adic metric d_3. Moreover, addition and multiplication in $\mathbb{Q}$ are continuous w.r.t. this metric, and the extensions of these operations let $\mathbb{Q}_3$ become a field.

Of special interest in our context is the subring $\mathbb{Z}_3 \subset \mathbb{Q}_3$ of 3-adic integers. We give a definition in terms of sequences of digits.

1.3. DEFINITION. The set of 3-*adic integers* is defined by

$$\mathbb{Z}_3 := \left\{ (a_j)_{j=0}^{\infty} : a_j \in \{0, 1, 2\} \text{ for each } j \in \mathbb{N}_0 \right\} \, .$$

The elements $a_j(x)$ of a 3-adic integer $x = (a_j(x))_{j=0}^{\infty} \in \mathbb{Z}_3$ are called 3-*adic digits* of x.

1.4. LEMMA. *The multiplicative group of invertible 3-adic integers is given by*

$$\mathbb{Z}_3^* = \left\{ (a_j)_{j=0}^\infty \in \mathbb{Z}_3 : a_0 \neq 0 \right\} .$$

If $(a_j)_{j=0}^\infty$ is a 3-adic integer, it is clear by Cauchy's criterion that the limit

(1.1) $$\sum_{j=0}^\infty a_j \cdot 3^j = \lim_{n \to \infty} \sum_{j=0}^{n-1} a_j \cdot 3^j$$

exists in the 3-adic metric. Moreover, in case the sequence (a_j) is eventually periodic, we can compute this limit using the geometric summation formula. This leads to some interesting embedding results.

1.5. LEMMA. *Let* $x = (a_j(x))_{j=0}^\infty \in \mathbb{Z}_3$. *Then*

(a)

$(a_j(x))_{j=0}^\infty$ *is eventually periodic* $\iff$ $x = r/s$ *with* $r, s \in \mathbb{Z}$ *and* $3 \nmid s$.

(b)

$a_j(x) = 0$ *for sufficiently large* j $\iff$ $x \in \mathbb{N}_0$.

(c)

$a_j(x) = 1$ *for sufficiently large* j $\iff$ $x \in \frac{1}{2} + \mathbb{Z}$.

(d)

$a_j(x) = 2$ *for sufficiently large* j $\iff$ $x \in -\mathbb{N}$.

Crucial in our context is the following extension of the notion of residue classes modulo 3^ℓ to 3-adic integers.

1.7. DEFINITION. Let $x, y \in \mathbb{Z}_3$ and $\ell \in \mathbb{N}$. We say that x *is congruent to* y *modulo* 3^ℓ, in symbols $x \equiv y \mod 3^\ell$, if $a_j(x) = a_j(y)$ for $j = 0, \ldots, \ell - 1$. A *residue class* modulo 3^ℓ is given by a set

$$\left\{ x \mod 3^\ell \right\} := \left\{ y \in \mathbb{Z}_3 : y \equiv x \mod 3^\ell \right\} .$$

Another notion which is very important in this context is the natural concept of integration on 3-adic numbers. Recall that a *left* (or *right*) *Haar* measure on a locally compact topological group is a positive Borel measure on that group which is invariant under left (or right) translations. Such a Haar measure exists, and is unique up to a constant factor. If the topological group under consideration is compact, left and right Haar measure coincide.*

In the present situation, we only have to consider compact groups. In this case, we shall require, in addition, that a Haar measure has total mass 1, and speak henceforth of *the* Haar measure of a compact group. We use the notation

$$\int_G f(a) \, da$$

to denote the integral w.r.t. the Haar measure of an integrable function f on a compact topological group G.

---

*A proof of existence of a doubly invariant measure on a compact topological group, based on the combinatorial concept of matching theory, can be found in [LoP].

1.8. LEMMA. *The groups $(\mathbb{Z}_3, +)$ and $(\mathbb{Z}_3^*, \cdot)$ are compact topological groups, where the topology is induced by the 3-adic metric. Moreover, the Haar measures μ_3 on $\mathbb{Z}_3$ and ν_3 on $\mathbb{Z}_3^*$ are given by*

$$\mu_3\left(\{x \bmod 3^\ell\}\right) = \frac{1}{3^\ell} \qquad and \qquad \nu_3\left(\{x \bmod 3^\ell\}\right) = \frac{1}{2 \cdot 3^{\ell-1}}$$

where $\{x \bmod 3^\ell\}$ denotes an arbitrary residue class in $\mathbb{Z}_3$ or $\mathbb{Z}_3^*$, respectively.

It is a nice fact that the restriction of the additive Haar measure of $\mathbb{Z}_3$ to the subset $\mathbb{Z}_3^*$ and the multiplicative Haar measure of the group $\mathbb{Z}_3^*$ are proportional to each other.

2. The estimating series

In this section, we use the inductive process described in corollary II.4.4 to extend the counting functions e_ℓ introduced in definition II.4.1 to the domain $\mathbb{N}_0 \times \mathbb{Q}_3$. Inserting these extended counting functions into the series in the estimate of theorem II.4.7, we are lead to a certain sequence of functions $s_n : \mathbb{Z}_3^* \to \overline{\mathbb{N}}_0$, where $\overline{\mathbb{N}}_0 := \mathbb{N}_0 \cup \{\infty\}$ denotes the extended set of non-negative integers. It is shown that the asymptotic behaviour of the functions s_n is connected to density estimates of $3n + 1$ predecessor sets. The section concludes with some facts about the values of the functions s_n.

Counting functions on 3-adic numbers. In fact, by corollary II.4.4, it is no problem to extend the counting functions e_ℓ to a larger domain of definition.

2.1. DEFINITION. *The sequence of extended counting function for admissible vectors*

$$e_\ell : \mathbb{N}_0 \times \mathbb{Q}_3 \to \mathbb{N}_0, \qquad for\ \ell \in \mathbb{N}_0,$$

is defined inductively by

(i) $$e_0(k, a) := \begin{cases} 1 & if\ a \in \mathbb{Z}_3 \\ 0 & if\ a \in \mathbb{Q}_3 \setminus \mathbb{Z}_3 \end{cases}$$

(ii) $$e_{\ell+1}(k, a) := \sum_{j=0}^{k} e_\ell\left(k - j, \frac{2^{j+1}a - 1}{3}\right) \qquad for\ \ell \in \mathbb{N}_0.$$

2.2. REMARK. We conclude from corollary II.4.4 that the functions $e_\ell : \mathbb{N}_0 \times \mathbb{Q}_3 \to \mathbb{N}_0$ just defined coincide on $\mathbb{N}_0 \times \mathbb{N} \subset \mathbb{N}_0 \times \mathbb{Q}_3$ with the counting functions of definition II.4.1.

It is easy to verify the properties of the counting functions e_ℓ given in lemma II.4.2 also for the extended counting functions defined above. For our purposes here, we record the analogues of parts (c) and (d) of that lemma.

2.3. LEMMA. *Let $\ell \in \mathbb{N}_0$. Then:*

(a) *If $\ell \geqslant 1$, the support of e_ℓ is contained in $\mathbb{N}_0 \times \mathbb{Z}_3^*$.*

(b) *If $a, b \in \mathbb{Z}_3$ with $a \equiv b \mod 3^\ell$, then $e_\ell(k, a) = e_\ell(k, b)$ for each $k \in \mathbb{N}_0$.*

PROOF. (Induction on ℓ.)

(a) It is clear, by definition, that $\mathrm{supp}(e_\ell) \subset \mathbb{N}_0 \times \mathbb{Z}_3$ for each $\ell \in \mathbb{N}_0$. From corollary 4.5 we know, in addition, that $\mathrm{supp}(e_1) \subset \mathbb{N}_0 \times \mathbb{Z}_3^*$.

Assume that $\mathrm{supp}(e_\ell) \subset \mathbb{N}_0 \times \mathbb{Z}_3^*$. Then the induction formula of definition 2.1 proves that $e_{\ell+1}(k, a)$ can be non-zero only if

$$\frac{2^{j+1}a - 1}{3} \in \mathbb{Z}_3^* \qquad \text{for some integer } j \text{ with } 0 \leqslant j \leqslant k.$$

This implies $2^{j+1}a - 1 \in \mathbb{Z}_3$ and $2^{j+1}a \equiv 1 \mod 3$, whence $2^{j+1}a \in \mathbb{Z}_3^*$. By lemma 1.4, we know that 2^{j+1} is invertible in $\mathbb{Z}_3$, hence $2^{-j-1} \in \mathbb{Z}_3^*$. We infer that

$$a = 2^{-j-1} \cdot 2^{j+1} a \in \mathbb{Z}_3^* \,.$$

This means, $e_{\ell+1}(k, a) \neq 0$ implies $a \in \mathbb{Z}_3^*$.

(b) For $\ell = 1$, the claim is clear by definition.

For the induction step, let $a, b \in \mathbb{Z}_3$ with $a \equiv b \mod 3^{\ell+1}$. Then we have

$$2^{j+1}a - 1 \equiv 2^{j+1}b - 1 \mod 3^{\ell+1} \qquad \text{for each } j \geqslant 0.$$

By part (a) and the induction formula in definition 2.1, we can restrict attention to those values of $j \in \mathbb{N}_0$ which render $2^{j+1}a \equiv 1 \mod 3$. Division by 3 gives

$$\frac{2^{j+1}a - 1}{3} \equiv \frac{2^{j+1}b - 1}{3} \mod 3^\ell \qquad \text{whenever } 2^{j+1}a \equiv 1 \mod 3,$$

which proves the claim using the induction hypothesis. □

The sequence of estimating series. Based on the series given in theorem II.4.7, we now introduce a sequence of functions on $\mathbb{Z}_3^*$ and investigate its relation to density estimates for the predecessor sets

$$\mathcal{P}_T(a) = \left\{ b \in \mathbb{N} : \text{there is a } k \in \mathbb{N}_0 \text{ with } T^k(b) = a \right\},$$

where $a \in \mathbb{N}$ and T is the $3n + 1$ function studied in chapter II.

2.4. DEFINITION. Let $\lambda := \log_2(\frac{3}{2}) \approx 0.58496$. For each $n \in \mathbb{N}$, the n-th *estimating series* is given by

$$s_n : \mathbb{Z}_3^* \to \overline{\mathbb{N}}_0, \qquad s_n(a) := \sum_{\ell=1}^{\infty} e_\ell(n + \lfloor \lambda\ell \rfloor, a) \,.$$

Note that the series starts with $\ell = 1$, because we prefer to have all the supports of the functions e_ℓ concentrated in $\mathbb{N}_0 \times \mathbb{Z}_3^*$. From a technical point of view, this occasionally will give some advantage.

It is clear that $s_n(a)$ is well-defined for each $a \in \mathbb{Z}_3^*$, even if the infinite sum is not bounded, because $e_\ell(n + \lfloor \alpha \ell \rfloor, a) \geq 0$ for each $n, \ell \in \mathbb{N}$ and $a \in \mathbb{Q}_3$. On the other hand, recall that $\mathcal{P}_T^n(a) = \{x \in \mathcal{P}_T(a) : 2^{n-1}a < x \leq 2^n a\}$. Then theorem II.4.7 gives the lower bound

$$(2.2) \qquad s_n(a) \leq |\mathcal{P}_T^n(a)| \leq 2^{n-1}a \qquad \text{if } a \in \mathbb{N} \text{ is non-cyclic;}$$

the upper bound is obvious, by the definition of $\mathcal{P}_T^n(a)$.

The asymptotic behaviour of $s_n(a)$ as $n \to \infty$ is connected to the 'size' of predecessor sets in the following way. We state the claim in the concise form of uniform lower bounds, compare remark II.6.2.

2.5. THEOREM. *Let $A \subset \mathbb{N}$, let $\beta \geq 1$ be a real number, and suppose that*

$$\liminf_{n \to \infty} \frac{s_n(a)}{\beta^n} > 0 \qquad \text{uniformly for } a \in A.$$

Then

$$\liminf_{x \to \infty} \left(\inf_{\substack{a \in A \text{ non-cyclic}}} \frac{Z_a(ax)}{x^{\log_2 \beta}} \right) > 0.$$

PROOF. The uniform lim inf-condition implies that there is a real number $\varrho > 0$ and an index $n_0 \in \mathbb{N}$ such that

$$s_n(a) \geq \varrho \, \beta^n \qquad \text{for } a \in A \text{ and } n \geq n_0.$$

This gives, using the bounds (2.2):

$$Z_a(2^n a) \geq |\mathcal{P}_T^n(a)| \geq \varrho \, \beta^n = \varrho \, e^{n \log \beta} \qquad \text{for non-cyclic } a \in A \text{ and } n \geq n_0.$$

Now let $x \in \mathbb{N}$ and put $n(x) := \lfloor \log_2 x \rfloor$. Then we obtain, for $x \geq 2^{n_0}$ and all non-cyclic $a \in A$:

$$\begin{aligned}
Z_a(ax) \geq Z_a\left(2^{n(x)}a\right) &\geq \varrho \, \exp(n(x) \log \beta) \\
&\geq \varrho \, \exp((\log_2 x - 1) \log \beta) \\
&= \varrho \, \exp(\log x \log_2 \beta - \log \beta) = \varrho \beta \, x^{\log_2 \beta} .
\end{aligned}$$

This completes the proof. □

2.6. REMARK. If $A = \mathbb{N} \backslash 3\mathbb{N}$, then lemma II.4.10 allows to omit the restriction to non-cyclic numbers. For general subsets $A \subset \mathbb{N}$, this argument is not valid. The condition that a should be non-cyclic comes in, as we make essential use of theorem II.4.7 which is only applicable to non-cyclic numbers $a \in \mathbb{N}$.

Ill-behaviour of the estimating series. In view of the bound (2.2) and theorem 2.5, the following result is a bit surprising.

2.7. THEOREM. *For each $n \in \mathbb{N}$, the niveau set $\{s_n = \infty\}$ is dense in $\mathbb{Z}_3^*$.*

PROOF. We show that for any given $a, \ell \in \mathbb{N}$ with $a \not\equiv 0 \mod 3$, there is an $x(a, \ell) \in \mathbb{Z}_3^*$ satisfying both

$$x(a, \ell) \equiv a \mod 3^\ell \qquad \text{and} \qquad s_n\big(x(a, \ell)\big) = +\infty \,.$$

This will prove the theorem, because the set of residue classes $\{a \mod 3^\ell\}$, where a, ℓ run through $\mathbb{N}$ with $a \not\equiv 0 \mod 3$, forms a basis for the topology on $\mathbb{Z}_3^*$ induced by the 3-adic metric.

It would be technically somewhat complicated to prove existence of $x(a, \ell)$ using only definitions 2.1 and 2.4, so we prefer to give a simpler proof based on the investigations of chapter II.

Let $(k_j)_{j \in \mathbb{N}}$ be a sequence of non-negative integers satisfying

$$(k_\ell, k_{\ell-1}, \ldots, k_0) \in \mathcal{E}(a) \,,$$

and

$$(2.3) \qquad k_0 + \cdots + k_j \leqslant n + \lfloor \lambda j \rfloor \qquad \text{for infinitely many } j \in \mathbb{N} \,.$$

(It is easy to fulfill these conditions; one could start with an arbitrary admissible vector w.r.t. a of length ℓ, which exists by lemma II.3.3, and put, for instance, $k_j = 0$ for $j > \ell$.) Now we construct $x(a, \ell)$ as the limit of a sequence $(x_j)_{j \in \mathbb{N}} \subset \mathbb{N}$ which converges in the 3-adic topology, and which fulfills

$$(2.4) \qquad\qquad (k_j, \ldots, k_0) \in \mathcal{E}(x_j) \,.$$

To start the inductive construction, put $x_j := a$ for $j = 1, \ldots, \ell$. Having defined x_j with the property (2.4), we infer from lemma II.3.2 (b) that there is a natural number x_{j+1} such that

$$x_{j+1} \equiv x_j \mod 3^j$$

and

$$(k_{j+1}, \ldots, k_0) = (k_{j+1}, 0) \cdot (k_j, \ldots, k_0) \in \mathcal{E}(x_{j+1}) \,.$$

Then the sequence $(x_j)_{j \in \mathbb{N}}$ converges in $\mathbb{Z}_3^*$, and the limit has the property

$$x(a, \ell) := \lim_{j \to \infty} x_j \equiv x_j \mod 3^j \qquad \text{for each } j \in \mathbb{N} \,.$$

Lemma 2.3 (b), relation (2.3) together with lemma II.4.2 (b), and (2.4) imply that

$$e_j\big(n + \lfloor \lambda j \rfloor, x(a, \ell)\big) = e_j\big(n + \lfloor \lambda j \rfloor, x_j\big) \geqslant e_j(k_0 + \cdots + k_j, x_j) \geqslant 1$$

for infinitely many $j \in \mathbb{N}$, which gives

$$s_n\big(x(a, \ell)\big) = \sum_{j=0}^{\infty} e_j\big(n + \lfloor \lambda j \rfloor, x(a, \ell)\big) = +\infty \,. \qquad \square$$

3. The averaged estimating series

Despite the fact that the estimating functions s_n of definition 2.4 behave so badly, as far as continuity properties are concerned (theorem 2.7), we shall see in this section that they are perfectly integrable and that their integrals have a nice form expressible in terms of binomial coefficients.

A formula for 3-adic averages. Recall definition 2.1 of the basic functions e_ℓ. We start out with their property of being constant on residue classes modulo 3^ℓ, observed in lemma 2.3, and recall lemma 1.8 concerning the normalized Haar measure on $\mathbb{Z}_3^*$.

3.1. DEFINITION. For $\ell \in \mathbb{N}$, denote by $A_\ell^* \subset \mathbb{N}$ a complete system of incongruent prime residues modulo 3^ℓ, e.g.

$$A_\ell^* = \left\{ 1, 2, 4, 5, \dots, 3^\ell - 2, 3^\ell - 1 \right\} .$$

The 3-*adic average* of e_ℓ is given by

$$\bar{e}_\ell : \mathbb{N}_0 \to \mathbb{Q}, \qquad \bar{e}_\ell(k) := \int_{\mathbb{Z}_3^*} e_\ell(k, a)\, da = \frac{1}{2 \cdot 3^{\ell-1}} \sum_{a \in A_\ell^*} e_\ell(k, a) .$$

One can easily compute the values of $\bar{e}_1$ from definition 2.1:

$$\bar{e}_1(k) = \frac{1}{2}\left(e_1(k, 1) + e_1(k, 2) \right) = \frac{1}{2}\left(\left\lfloor \frac{k+1}{2} \right\rfloor + \left\lceil \frac{k+1}{2} \right\rceil \right) = \frac{k+1}{2} .$$

This extends to a general formula for all 3-adic averages.

3.2. LEMMA. *For any integers $k \geqslant 0$ and $\ell \geqslant 1$ we have*

$$\bar{e}_\ell(k) = \frac{1}{2 \cdot 3^{\ell-1}} \binom{k+\ell}{\ell} .$$

PROOF. The proof is based on the induction rule of definition 2.1,

$$e_{\ell+1}(k, a) = \sum_{j=0}^{k} e_\ell\left(k - j, \frac{2^{j+1}a - 1}{3} \right) .$$

Let us fix j, and suppose that a runs through the set $A_{\ell+1}^*$. Then the integer values of the fraction

$$\frac{2^{j+1}a - 1}{3}$$

run exactly once through a complete system A_ℓ of incongruent prime residues modulo 3^ℓ. Hence,

$$\bar{e}_{\ell+1}(k) = \frac{1}{2 \cdot 3^\ell} \sum_{a \in A_{\ell+1}^*} e_{\ell+1}(k, a) = \frac{1}{2 \cdot 3^\ell} \sum_{a \in A_{\ell+1}^*} \sum_{j=0}^{k} e_\ell \left(k - j, \frac{2^{j+1}a - 1}{3} \right)$$

$$= \frac{1}{2 \cdot 3^\ell} \sum_{j=0}^{k} \sum_{b \in A_\ell} e_\ell(k - j, b) = \frac{1}{3} \sum_{j=0}^{k} \frac{1}{2 \cdot 3^{\ell-1}} \sum_{b \in A_\ell^*} e_\ell(k - j, b)$$

$$= \frac{1}{3} \sum_{j=0}^{k} \bar{e}_\ell(j) .$$

Taking into account the values for $\bar{e}_1(k)$ above, the result is implied by the well-known formula (see, for instance, [Fel], chap. II., problem 12.7)

$$\sum_{j=0}^{k} \binom{j + \ell}{\ell} = \binom{k + \ell + 1}{\ell + 1} . \quad \square$$

The averaged estimating series. Based, essentially, on Stirling's formulaand Beppo Levi's theorem, it is now possible to show that the functions s_n defined in 2.4 are integrable.

3.3. NOTATION. For $\ell, n \in \mathbb{N}$, we shall use the notation

$$\beta_{\ell,n} := \bar{e}_\ell(n + \lfloor \lambda \ell \rfloor) = \frac{1}{2 \cdot 3^{\ell-1}} \binom{n + \lfloor \alpha \ell \rfloor}{\ell}$$

with $\alpha = \log_2 3 \approx 1.58496$ and $\lambda = \log_2(\frac{3}{2}) = \alpha - 1$.

We are now heading to prove $\sum_{\ell=1}^{\infty} \beta_{\ell,n} < \infty$. In view of definition 2.4 and B. Levi's theorem, this will imply both that the s_n are integrable, and the central equality in

$$\int_{\mathbb{Z}_3^*} s_n(a) \, da = \int_{\mathbb{Z}_3^*} \sum_{\ell=1}^{\infty} e_\ell(n + \lfloor \lambda \ell \rfloor, a) \, da = \sum_{\ell=1}^{\infty} \int_{\mathbb{Z}_3^*} e_\ell(n + \lfloor \lambda \ell \rfloor, a) \, da = \sum_{\ell=1}^{\infty} \beta_{\ell,n} .$$

3.4. LEMMA. *Suppose that n, k are integers satisfying $0 < k < n$, denote $r := k/n$, and use the function $h(x) = x \log x + (1 - x) \log(1 - x)$ for $0 < x < 1$. Then the number $c(n, r)$ defined by*

$$\binom{n}{k} = c(n, r) \, (2\pi n r(1 - r))^{-1/2} \exp(-nh(r))$$

fulfills the estimates

$$\exp \left(-\frac{1}{12nr(1 - r)} \right) < c(n, r) < \exp \left(\frac{1}{12n} \right) .$$

PROOF. This is a consequence of Stirling's estimate, cf. [Fel], pp. 50ff and 169f. $\square$

3.5. COROLLARY. *Let* $(n_\nu)_{\nu\in\mathbb{N}}, (k_\nu)_{\nu\in\mathbb{N}} \subset \mathbb{N}$ *be two sequences, and suppose that* $\lim\limits_{\nu\to\infty} n_\nu = \infty$ *and that the* $r_\nu := k_\nu/n_\nu$ *converge to a limit* r *with* $0 < r < 1$. *Then, with the notation of the preceding lemma,* $\lim\limits_{\nu\to\infty} c(n_\nu, r_\nu) = 1$.

3.6. LEMMA. *Let* $n \in \mathbb{N}$. *Then* $\sum_{\ell=1}^{\infty} \beta_{\ell,n}$ *converges.*

PROOF. Put $n_\ell := n + \lfloor \alpha\ell \rfloor$ and $k_\ell := \ell$, for $\ell \in \mathbb{N}$. Then

$$r_\ell = \frac{k_\ell}{n_\ell} = \frac{\ell}{n + \lfloor \alpha\ell \rfloor} \longrightarrow \frac{1}{\alpha} \qquad \text{as } \ell \to \infty.$$

According to corollary 3.5 we know that

$$\binom{n + \lfloor \alpha\ell \rfloor}{\ell} = c(n_\ell, r_\ell) \left(\frac{2\pi n_\ell}{\alpha} \left(1 - \frac{1}{\alpha} \right) \right)^{-1/2} \exp\left(-n_\ell\, h\left(\frac{1}{\alpha} \right) \right),$$

where $\lim\limits_{\ell\to\infty} c(n_\ell, r_\ell) = 1$. Now convergence is proved by the root criterion,

$$\begin{aligned}
\lim_{\ell\to\infty} \sqrt[\ell]{\beta_{\ell,n}} &= \lim_{\ell\to\infty} \left(\frac{c(n_\ell, r_\ell)}{2 \cdot 3^{\ell-1}} \right)^{1/\ell} \left(\frac{2\pi n_\ell}{\alpha} \left(1 - \frac{1}{\alpha} \right) \right)^{-1/(2\ell)} \exp\left(-\frac{n_\ell}{\ell} h\left(\frac{1}{\alpha} \right) \right) \\
&= \frac{1}{3} \exp\left(-\alpha\, h\left(\frac{1}{\alpha} \right) \right) \\
&= \frac{1}{3} \exp\left(-\log\frac{1}{\alpha} - (\lambda) \log\left(1 - \frac{1}{\alpha} \right) \right) \\
&= \frac{\alpha}{3} \left(1 - \frac{1}{\alpha} \right)^{-\lambda} = \frac{\alpha}{3} \left(1 + \frac{1}{\lambda} \right)^{\lambda} \approx 0.94650 < 1. \quad \square
\end{aligned}$$

3.7. COROLLARY. *Let* $n \in \mathbb{N}$. *Then* $s_n : \mathbb{Z}_3^* \to \mathbb{R}$ *is integrable w.r.t. the Haar measure on* $\mathbb{Z}_3^*$, *and*

$$\int_{\mathbb{Z}_3^*} s_n(a)\, da = \sum_{\ell=1}^{\infty} \beta_{\ell,n}.$$

3.8. DEFINITION. *The* averaged estimating sums *are defined by*

$$\bar{s}_n := \int_{\mathbb{Z}_3^*} s_n(a)\, da = \sum_{\ell=1}^{\infty} \beta_{\ell,n} \qquad \text{for } n \in \mathbb{N}.$$

4. Maximal terms

This section is devoted to the question: which is the maximal term in the averaged estimating sum (remember $\alpha = \log_2 3 \approx 1.58496$)

$$(4.1) \qquad \bar{s}_n = \sum_{\ell=1}^{\infty} \beta_{\ell,n} = \sum_{\ell=1}^{\infty} \frac{1}{2 \cdot 3^{\ell-1}} \binom{n + \lfloor \alpha\ell \rfloor}{\ell} \quad ?$$

The main ingredients in that series are binomial coefficients, and it is well-known that, for fixed $n \in \mathbb{N}$, the binomial coefficients $\binom{n}{k}$ reach their maximum when $k = \lfloor n/2 \rfloor$ or $k = \lceil n/2 \rceil$. Does this fact give a hint at which index $\ell = \ell_0(n)$, for fixed $n \in \mathbb{N}$, the $\beta_{\ell,n}$ assume their maximum? On a first look: no, because the $\beta_{\ell,n}$ are not just binomial coefficients with fixed upper entry. Indeed, the upper entry $n + \lfloor \alpha \ell \rfloor$ is increasing with ℓ. But, on the other hand, there is a weight $(2 \cdot 3^{\ell-1})^{-1}$ in front of the binomial coefficient which compensates for the increase in the upper entry. It will turn out that there is a sequence of indices $\ell_0(n)$ such that

$$(4.2) \qquad 2\,\ell_0(n) = n + \lfloor \alpha \ell_0(n) \rfloor \qquad \text{for each } n \in \mathbb{N},$$

and, moreover, that this index is—*asymptotically*—a good candidate for the index of the maximal term in the sum (4.1).

There is yet another, and more interesting, heuristic argument for $\beta_{\ell_0(n),n}$ to be, approximately, the maximal term in the sum giving $\overline{s}_n$. Recall the formulae of notation 3.3 and definition 3.1,

$$\beta_{\ell,n} = \overline{e}_\ell(n + \lfloor \lambda \ell \rfloor) = \frac{1}{2 \cdot 3^{\ell-1}} \sum_{a \in A_\ell^*} e_\ell(n + \lfloor \lambda \ell \rfloor, a),$$

with $\lambda = \log_2(\tfrac{3}{2}) = \alpha - 1$. The quantities $e_\ell(k, a)$ occuring in the last sum originally had been defined as counting functions for admissible vectors w.r.t. $a \in \mathbb{N}$ with length ℓ and absolute value k (see definition II.4.1). Now, by theorem II.2.18, a vector of the type counted in $e_\ell(k, a)$ corresponds to a path in the Collatz graph terminating at the vertex $a \in V_T = \mathbb{N}$ and consisting of precisely k edges arising from T_0 and precisely ℓ edges arising from T_1, or to a parity vector (definition II.2.1) with k 0's and ℓ 1's. Now the usual $3n + 1$ heuristics (cf., for instance, J. C. Lagarias [**Lag1**], or S. Wagon [**Wag**]), asserts that most parity vectors have, roughly, as much 0's as 1's. Hence, we should expect that the $e_\ell(k, a)$ are largest if $k \approx \ell$, or, in terms of $\beta_{\ell,n}$, if

$$\ell \approx n + \lfloor \lambda \ell \rfloor,$$

a condition which is best verified by the index $\ell_0(n)$ of (4.2) (because $\lambda = \alpha - 1$).

The candidate for the maximal term. We are now going to give some explicit estimates concerning the supposed maximal terms in the series (4.1). This is, in fact, not very difficult after lemma 3.4.

4.1. LEMMA. *Let* $\lambda = \log_2(\tfrac{3}{2})$, *and put* $\ell_0(n) := \left\lfloor \dfrac{n}{1 - \lambda} \right\rfloor$, *for* $n \in \mathbb{N}$. *Then we have*

(a) $\ell_0(n) = n + \lfloor \lambda \ell_0(n) \rfloor$ *for each* $n \in \mathbb{N}$.
(b) *The constants* $\tilde{c}_0(n)$ *defined by*

$$\beta_{\ell_0(n),n} = \tilde{c}_0(n) \cdot \frac{3}{2} \left(\frac{1 - \lambda}{\pi} \right)^{1/2} \frac{2^n}{\sqrt{n}}$$

satisfy the estimate

$$\frac{3}{4} \exp\left(\frac{\lambda - 1}{6(n - 1 + \lambda)}\right) < \tilde{c}_0(n) < \left(1 + \frac{1 - \lambda}{2(n - 1 + \lambda)}\right) \exp\left(\frac{1 - \lambda}{24n}\right).$$

PROOF. (a) Let's calculate:

(4.3)
$$\ell_0(n) := \left\lfloor \frac{n}{1 - \lambda} \right\rfloor \iff \frac{n - 1 + \lambda}{1 - \lambda} < \ell_0(n) \leqslant \frac{n}{1 - \lambda}$$

(4.4)
$$\iff \ell_0(n) \leqslant n + \lambda\ell_0(n) < \ell_0(n) + (1 - \lambda).$$

As $\lambda \approx 0.58496$, we have $0 < 1 - \lambda < 1$, and this implies (a).

(b) In order to apply lemma 3.4, we use the notations $r := \ell/(n + \lfloor \alpha\ell \rfloor)$ and $h(r) = r \log r + (1 - r) \log(1 - r)$, to obtain

(4.5)
$$\binom{n + \lfloor \alpha\ell \rfloor}{\ell} = c_1 \cdot (2\pi(n + \lfloor \alpha\ell \rfloor)r(1 - r))^{-1/2} \exp\left(-(n + \lfloor \alpha\ell \rfloor) h(r)\right),$$

where the constant c_1 satisfies

(4.6)
$$\exp\left(-\frac{1}{12(n + \lfloor \alpha\ell \rfloor)r(1 - r)}\right) < c_1 < \exp\left(\frac{1}{12(n + \lfloor \alpha\ell \rfloor)}\right).$$

Now put $\ell := \ell_0(n)$. Then (a) gives, using $\alpha = \lambda + 1$,

$$n + \lfloor \alpha\ell_0(n) \rfloor = n + \lfloor \lambda\ell_0(n) \rfloor + \ell_0(n) = 2\,\ell_0(n),$$

from which we infer $r = \frac{1}{2}$ and $h(r) = -\log 2$. Inserting this into (4.5) gives

$$\binom{n + \lfloor \alpha\ell_0(n) \rfloor}{\ell} = c_1 \cdot c_2 \cdot \left(\frac{1 - \lambda}{\pi}\right)^{1/2} \cdot c_3 \cdot 2^{n + \alpha\ell_0(n)},$$

with the additional constants

$$c_2 := \left(\frac{n}{(1 - \lambda)\ell_0(n)}\right)^{1/2},$$

$$c_3 := \exp\left((\lfloor \alpha\ell_0(n) \rfloor - \alpha\ell_0(n)) \log 2\right) = 2^{\lfloor \lambda\ell_0(n) \rfloor - \lambda\ell_0(n)}.$$

To estimate c_2, we use (4.3) to obtain

$$1 \leqslant c_2 < \left(\frac{n}{n - 1 + \lambda}\right)^{1/2} < 1 + \frac{1 - \lambda}{2(n - 1 + \lambda)}.$$

Concerning c_3, the inequalities (4.4) yield

$$\lambda - 1 < \lfloor \lambda\ell_0(n) \rfloor - \lambda\ell_0(n) \leqslant 0, \quad \text{whence} \quad \frac{3}{4} < c_3 \leqslant 1.$$

In addition, (4.3) and part (a) imply the inequalities

$$\frac{n - 1 + \lambda}{1 - \lambda} < \frac{n + \lfloor \alpha \ell_0(n) \rfloor}{2} \leqslant \frac{n}{1 - \lambda}.$$

This gives, together with $r = \frac{1}{2}$ and (4.6),

$$\exp\left(\frac{\lambda - 1}{6(n - 1 + \lambda)}\right) < c_1 < \exp\left(\frac{1 - \lambda}{24n}\right).$$

Summarizing, we have

$$\beta_{\ell_0(n), n} = \frac{1}{2 \cdot 3^{\ell_0(n) - 1}} \binom{2\ell_0(n)}{\ell_0(n)} = \tilde{c}_0(n) \cdot \frac{3}{2}\left(\frac{1 - \lambda}{\pi}\right)^{1/2} \frac{2^n}{\sqrt{n}}$$

with $\tilde{c}_0(n) := c_1 c_2 c_3$ satisfying

$$\frac{3}{4}\exp\left(\frac{\lambda - 1}{6(n - 1 + \lambda)}\right) < \tilde{c}_0(n) < \left(1 + \frac{1 - \lambda}{2(n - 1 + \lambda)}\right)\exp\left(\frac{1 - \lambda}{24n}\right). \quad \square$$

An estimate for the remaining terms.   In order to estimate generally the terms $\beta_{\ell,n}$, we employ again lemma 3.4.

4.2. LEMMA. *Let* $\alpha = \lambda + 1 = \log_2 3$, *let* $\ell, n \in \mathbb{N}$, *and put* $r := \ell/(n + \lfloor \alpha\ell \rfloor)$. *Then the constants* $\tilde{c}(\ell, r)$ *defined by*

$$\beta_{\ell,n} = \tilde{c}(\ell, r) \cdot \frac{3}{2}\left(2\pi\ell(1 - r)\right)^{-1/2} \exp\left(-nh(r) + \alpha\ell\left(h\left(\tfrac{1}{2}\right) - h(r)\right)\right)$$

satisfy the estimate

$$\frac{1}{2}\exp\left(\frac{-1}{12\ell(1 - r)}\right) < \tilde{c}(\ell, r) < \exp\left(\frac{r}{12\ell}\right).$$

PROOF. As in the proof of the previous lemma, we start with a form of (4.5),

$$(4.7) \qquad \binom{n + \lfloor \alpha\ell \rfloor}{\ell} = c_1 \cdot \left(2\pi\ell(1 - r)\right)^{-1/2} \exp\left(-(n + \lfloor \alpha\ell \rfloor) h(r)\right).$$

Here the constant c_1 satisfies

$$(4.8) \qquad \exp\left(-\frac{1}{12\ell(1 - r)}\right) < c_1 < \exp\left(\frac{r}{12\ell}\right).$$

To do the next step towards an estimate for $\beta_{\ell,n}$, we multiply (4.7) by $3^{-\ell} = 2^{-\alpha\ell} = \exp\left(\alpha\ell\, h(\tfrac{1}{2})\right)$ and treat first the resulting argument of the exp term:

$$-(n + \lfloor \alpha\ell \rfloor) h(r) - \alpha\ell\, h\left(\tfrac{1}{2}\right) = -n\, h(r) + \alpha\ell\left(h\left(\tfrac{1}{2}\right) - h(r)\right) + (\alpha\ell - \lfloor \alpha\ell \rfloor) h(r).$$

This gives

$$\frac{1}{3^\ell}\binom{n+\lfloor\alpha\ell\rfloor}{\ell} = c_1 \cdot \left(2\pi\ell(1-r)\right)^{-1/2} \cdot c_4 \cdot \exp\left(-n\,h(r) + \alpha\ell\left(h\left(\tfrac{1}{2}\right) - h(r)\right)\right),$$

with the additional constant $c_4 = \exp\left((\alpha\ell - \lfloor\alpha\ell\rfloor)h(r)\right)$. The function $h(x)$ assumes its minimum at $x = \frac{1}{2}$, and satisfies $-\log 2 \leqslant h(x) < 0$ for $0 < x < 1$. From this, we conclude

$$-\log 2 < (\alpha\ell - \lfloor\alpha\ell\rfloor)h(r) < 0, \qquad \text{whence} \qquad \frac{1}{2} < c_4 \leqslant 1.$$

Summarizing, this implies together with (4.8) that

$$\beta_{\ell,n} = \frac{1}{2\cdot 3^{\ell-1}}\binom{n+\lfloor\alpha\ell\rfloor}{\ell} = \tilde{c}(\ell,r)\cdot\frac{3}{2}\left(2\pi\ell(1-r)\right)^{-1/2}\exp\left(-(n+\lfloor\alpha\ell\rfloor)\,h(r)\right),$$

with $\tilde{c}(\ell,r) := c_1 c_4$ satisfying

$$\frac{1}{2}\exp\left(-\frac{1}{12\ell(1-r)}\right) < \tilde{c}(\ell,r) < \exp\left(\frac{r}{12\ell}\right). \quad \square$$

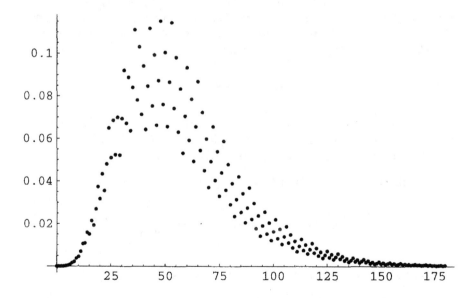

FIGURE 2. This figure shows the values of $\beta_{\ell,n}$ for $n = 20$ and $\ell = 1, \ldots, 180$.

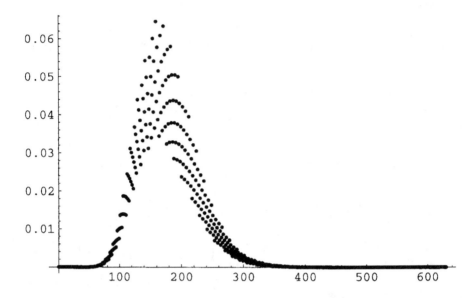

FIGURE 3. This figure shows the values of $\beta_{\ell,n}$ for $n = 70$ and $\ell = 1, \ldots, 630$.

First order asymptotics of maximal terms. Observe that the estimates of lemmas 4.1 and 4.2 for the constants $\tilde{c}_0(n)$ and $\tilde{c}(\ell, r)$ contain factors $3/4$ and $1/2$, respectively. This makes it impossible to determine precisely the asymptotic behaviour of $\beta_{\ell_0(n),n}$ as $n \to \infty$, or that of $\beta_{\ell,n}$ as ℓ and n both tend to ∞, using only those estimates. Moreover, they seem to be not appropriate to determine *precisely* the location $\ell_1(n)$ of a maximal term

$$(4.9) \qquad\qquad \beta_{\ell_1(n),n} = \max_{\ell \in \mathbb{N}} \beta_{\ell,n}$$

for any given $n \in \mathbb{N}$ (or even for sufficiently large n). On the other hand, the estimates of lemmas 4.1 and 4.2 suffice to determine the first order term of the asymptotic behaviour of a funciton $\ell_1(n)$ satisfying (4.9).

4.3. THEOREM. *Let* $\ell_1 : \mathbb{N} \to \mathbb{N}$ *satisfy* (4.9). *Then*

$$\lim_{n \to \infty} \frac{\ell_1(n)}{n} = \lim_{n \to \infty} \frac{\ell_0(n)}{n} = \frac{1}{1 - \lambda} = \frac{1}{2 - \log_2 3} \approx 2.40942 \,.$$

PROOF. Let's start with the following estimates

$$(4.10) \qquad\qquad 0 < r_n := \frac{\ell_1(n)}{n + \lfloor \alpha \ell_1(n) \rfloor} < \frac{1}{\alpha} < 1,$$

which are valid for each $n \in \mathbb{N}$. We are looking for facts about the limiting behaviour of the quotients

$$Q_n := \frac{\beta_{\ell_1(n),n}}{\beta_{\ell_0(n),n}}, \qquad \text{as } n \to \infty.$$

Note that condition (4.9) on $\ell_1(n)$ implies $Q_n \geqslant 1$ for each $n \in \mathbb{N}$. Using lemmas 4.1 and 4.2, we estimate these quotients

$Q_n =$

$$= \frac{\tilde{c}(\ell_1(n), r_n)}{\tilde{c}_0(n)} \left(\frac{n}{2\ell_1(n)(1-r_n)(1-\lambda)} \right)^{1/2} \exp\left((n + \alpha\ell_1(n)) \left(h(\tfrac{1}{2}) - h(r_n) \right) \right)$$

$$< \frac{\tfrac{4}{3} \exp\left(\frac{r_n}{12\ell_1(n)} - \frac{\lambda-1}{6(n-1+\lambda)} \right)}{\left(2(1-r_n)(1-\lambda) \right)^{1/2}} \exp\left((n + \alpha\ell_1(n)) \left(h(\tfrac{1}{2}) - h(r_n) \right) + \tfrac{1}{2} \log n \right);$$

in the last expression, concerning the argument of the log, we also used the simple fact that $n/\ell_1(n) \leqslant n$.

Applying the upper estimate of (4.10), it is easily seen that the first factor in the last expression is bounded above. That means, in order to see what can happen with Q_n as $n \to \infty$, we have to investigate the limiting behaviour of

$$Y_n := (n + \alpha\ell_1(n)) \left(h(\tfrac{1}{2}) - h(r_n) \right) + \tfrac{1}{2} \log n, \qquad \text{as } n \to \infty.$$

We already observed (in the proof of lemma 4.2) that the function $h(r)$ assumes its minimum only at $r = \tfrac{1}{2}$, from which we obtain $h(\tfrac{1}{2}) - h(r_n) < 0$ whenever $r_n \neq \tfrac{1}{2}$. If the sequence (r_n) happens to have a subsequence $(r_{n_k})_{k \in \mathbb{N}}$ which converges to some $\gamma \neq \tfrac{1}{2}$, then we obtain that the subsequence Y_{n_k} tends to $-\infty$, as $k \to \infty$. But this would have the consequence

$$\lim_{k \to \infty} Q_{n_k} = 0,$$

which would be a contradiction to condition (4.9) on $\ell_1(n)$. $\square$

5. Asymptotic behaviour of the averaged sums

It is proved in this section that the averaged estimating sums have the significant property that

(5.1) $$\liminf_{n \to \infty} \frac{\bar{s}_n}{2^n} > 0.$$

Suppose that we could prove, for some non-cyclic $a \in \mathbb{N}$, that $s_n(a)$ is very close to $\bar{s}_n$, if n is large. Then (5.1) and theorem 2.5 would imply that the predecessor set $\mathcal{P}_T(a)$ has positive asymptotic density, in the sense that there is a constant $c > 0$ such that

$$Z_{\mathcal{P}_T(a)}(x) = \left| \{ \nu \in \mathcal{P}_T(a) : \nu \leqslant x \} \right| \geqslant c \, x \qquad \text{for sufficiently large } n \in \mathbb{N}.$$

This would be a very deep property of the Collatz graph of the $3n+1$ function.

The purpose of this section is to prove (5.1). Unfortunately, to do this, we need a complicated analysis of the $\beta_{\ell,n}$—even more complicated than that given in the previous section.

A naive approach.   In order to provide a better motivation for this analysis, let us see what happens if we take a more naive approach using a simple binomial identity.

$$\beta_{\ell,n+1} = \frac{1}{2 \cdot 3^{\ell-1}} \binom{n+1+\lfloor \alpha\ell \rfloor}{\ell} = \frac{1}{2 \cdot 3^{\ell-1}} \left(\binom{n+\lfloor \alpha\ell \rfloor}{\ell} + \binom{n+\lfloor \alpha\ell \rfloor}{\ell-1} \right)$$

(5.2)

$$= \beta_{\ell,n} + \frac{1}{3} \cdot \frac{1}{2 \cdot 3^{\ell-2}} \binom{n+\lfloor \alpha(\ell-1)+\alpha \rfloor}{\ell-1} \geqslant \beta_{\ell,n} + \frac{1}{3} \beta_{\ell-1,n+1}$$

Here we used $\alpha > 1$ which yields

(5.3) $$n + \lfloor \alpha(\ell-1) + \alpha \rfloor \geqslant n + 1 + \lfloor \alpha(\ell-1) \rfloor \,.$$

From (5.2), we obtain $\beta_{\ell,n+1} - \frac{1}{3}\beta_{\ell-1,n+1} \geqslant \beta_{\ell,n}$. Summing over $\ell \in \mathbb{N}$ (with $\beta_{0,n+1} = \frac{3}{2}$) gives

$$\bar{s}_{n+1} - \tfrac{1}{3}\left(\bar{s}_{n+1} + \tfrac{3}{2}\right) \geqslant \bar{s}_n \,, \qquad \text{or} \qquad \bar{s}_{n+1} \geqslant \tfrac{3}{2}\left(\bar{s}_n + \tfrac{1}{2}\right) \,,$$

which only implies

$$\liminf_{n \to \infty} \frac{\bar{s}_n}{\beta^n} > 0 \qquad \text{with} \quad \beta = \frac{3}{2} \,.$$

This is really much worse than (5.1). The large approximation error creeped in in (5.3); indeed, we have

$$n + \lfloor \alpha(\ell-1) + \alpha \rfloor = \begin{cases} n + 1 + \lfloor \alpha(\ell-1) \rfloor & \text{or} \\ n + 2 + \lfloor \alpha(\ell-1) \rfloor \,, \end{cases}$$

where the two branches occur in a complicated alternation determined by the irrational number $\alpha = \log_2 3$.

The theorem.   A better method to estimate $\bar{s}_n$ relies on the approximation of the binomial coefficients by the Gaußian bell-shaped function.

5.1. LEMMA. *Suppose that n, k are integers satisfying $0 < k < n$, denote $r := k/n$ and $x := \sqrt{n}(2r - 1)$. Then the numbers $c^*(n, x)$ defined by*

$$\frac{1}{2^n} \binom{n}{k} = c^*(n, x) \left(\frac{2}{\pi n}\right)^{1/2} \exp\left(-\frac{x^2}{2}\right)$$

are given by

$$c^*(n, x) = \frac{c(n, r)}{2\sqrt{r(1-r)}} \cdot \exp\left(-x^2 \, F\left(\frac{x}{\sqrt{n}}\right)\right) \,,$$

where $c(n, r)$ is taken from lemma 3.4, and

$$F(\xi) := \frac{1+\xi}{2\xi^2} \log(1+\xi) + \frac{1-\xi}{2\xi^2} \log(1-\xi) - \frac{1}{2} = \sum_{m=1}^{\infty} \frac{\xi^{2m}}{(2m+1)(2m+2)}$$

satisfies $\lim_{\xi \to 0} F(\xi) = 0$.

PROOF. This is, in fact, a corollary of lemma 3.4 obtained using Taylor expansion fo the function $h(x) = x \log x + (1-x) \log(1-x)$, cf. [**Fel**], p. 169f.   □

5.2. THEOREM. $\displaystyle\liminf_{n\to\infty} \frac{\bar{s}_n}{2^n} > 0$.

PROOF. We shall use the preceding lemma to estimate sufficiently many terms in the series

$$(5.4) \qquad \frac{\bar{s}_n}{2^n} = \sum_{\ell=1}^{\infty} \frac{\beta_{\ell,n}}{2^n} = \frac{3}{2} \sum_{\ell=1}^{\infty} \frac{1}{2^n 3^\ell} \binom{n + \lfloor \alpha\ell \rfloor}{\ell} \geq \frac{3}{4} \sum_{\ell=1}^{\infty} \frac{1}{2^{n_\ell}} \binom{n_\ell}{\ell},$$

where we used the notation $n_\ell := n + \lfloor \alpha\ell \rfloor$. Note the estimates

$$(5.5) \qquad n + \alpha\ell - 1 < n_\ell \leq n + \alpha\ell.$$

The aim is to prove that there is an overall constant $c > 0$ such that $2^{-n}\bar{s}_n \geq c$ for sufficiently large $n \in \mathbb{N}$. To this end, we employ lemma 5.1 which asserts that, if $c(\ell)$ is defined by

$$(5.6) \qquad \frac{1}{2^{n_\ell}} \binom{n_\ell}{\ell} = c(\ell) \cdot \left(\frac{2}{\pi n_\ell}\right)^{1/2} \exp\left(-\frac{x_\ell^2}{2}\right) \qquad \text{with} \quad x_\ell = \frac{2\ell - n_\ell}{\sqrt{n_\ell}},$$

and if, in addition, $|x_\ell| \leq 1$ for each $\ell \in \mathbb{N}$, then

$$(5.7) \qquad \lim_{\ell\to\infty} c(\ell) = 1.$$

Now the method is to ask for the range of ℓ such that $x_\ell^2 \leq 1$, for each $n \in \mathbb{N}$. This means, we shall seek a sequence of subsets

$$\Delta_n \subset \{\ell \in \mathbb{N} : x_\ell^2 \leq 1\}$$

satisfying the two conditions

$$(5.8) \qquad \lim_{n\to\infty} (\inf \Delta_n) = \infty \qquad \text{and} \qquad |\Delta_n| \text{ is sufficiently large,}$$

which will lead through (5.6) and (5.7) to an estimate for (5.4).

Let's fix $n \in \mathbb{N}$ and start the calculation.

$$(5.9) \qquad x_\ell^2 \leq 1 \quad\Longleftrightarrow\quad (2\ell - n_\ell)^2 \leq n_\ell.$$

First we want to get rid of the floor function involved in n_ℓ. This will be achieved by the estimates $n + \alpha\ell - 1 < n_\ell \leq n + \alpha\ell$, which show that $2\ell - n - \alpha\ell \leq 2\ell - n_\ell < 2\ell - n - \alpha\ell + 1$ and

$$(2\ell - n_\ell)^2 \leq \left((2 - \alpha)\ell - n\right)^2 + \max\{0, 1 + 2(2 - \alpha)\ell - 2n\}.$$

It will suffice for us to restrict attention to those ℓ which render the last maximum equal to 0. Formally,

$$(5.10) \qquad \ell \leq \frac{n}{2 - \alpha} \qquad \text{and} \qquad \left((2 - \alpha)\ell - n\right)^2 \leq n + \alpha\ell - 1$$

clearly ensure the second estimate of (5.9), and hence $x_\ell^2 \leqslant 1$ for such ℓ.

The second estimate of (5.10) amounts to

$$(5.11) \qquad (2-\alpha)^2 \ell^2 - (2(2-\alpha)n + \alpha)\ell + n^2 - n + 1 \leqslant 0.$$

This is a quadratic inequality with discriminant

$$D_n := (2(2-\alpha)n + \alpha)^2 - 4(2-\alpha)^2(n^2 - n + 1) = 8(2-\alpha)n + \alpha^2 - 4(2-\alpha)^2.$$

As $\alpha \approx 1.5$, the coefficient of n is positive, and the constant term is $\alpha^2 - 4(2 - \alpha)^2 \approx 1.3$. Hence, there are two constants $c_2 > c_1 > 0$ such that

$$(5.12) \qquad c_1 n \leqslant D_n \leqslant c_2 n \qquad \text{for} \quad n \in \mathbb{N}.$$

With the smaller solution of (5.11),

$$(5.13) \qquad \ell_-(n) := \frac{n + \alpha}{2 - \alpha} - \frac{\sqrt{D_n}}{2(2-\alpha)^2},$$

let us put

$$(5.14) \qquad \Delta_n := \left\{ \lceil \ell_-(n) \rceil, \ldots, \left\lfloor \frac{n}{2-\alpha} \right\rfloor \right\} \subset \{\ell \in \mathbb{N} : x_\ell^2 \leqslant 1\}.$$

According to (5.8), we have to estimate both $\inf \Delta_n$ and $|\Delta_n|$. But (5.12) and (5.13) show that

$$\left\lfloor \frac{n}{2-\alpha} \right\rfloor - c_3\sqrt{n} \leqslant \lceil \ell_-(n) \rceil \leqslant \left\lfloor \frac{n}{2-\alpha} \right\rfloor - c_4\sqrt{n}$$

for appropriate constants $c_3 > c_4 > 0$ and large n. From this we infer both

$$(5.15) \qquad \lim_{n \to \infty} (\inf \Delta_n) = \infty \qquad \text{and}$$

$$(5.16) \qquad |\Delta_n| \geqslant c_4\sqrt{n} \qquad \text{for large } n.$$

Moreover, (5.14) and the second estimate of (5.5) yield

$$(5.17) \qquad n_\ell \leqslant n + \alpha \left\lfloor \frac{n}{2-\alpha} \right\rfloor \leqslant \frac{2}{2-\alpha} n \qquad \text{for } \ell \in \Delta_n.$$

Now we are ready to do the final calculation

$$\frac{\bar{s}_n}{2^n} \geqslant \frac{3}{4} \sum_{\ell=1}^{\infty} \frac{1}{2^{n_\ell}} \binom{n_\ell}{\ell} \qquad \qquad \text{by (5.4),}$$

$$= \frac{3}{4} \sum_{\ell=1}^{\infty} c(\ell) \cdot \left(\frac{2}{\pi n_\ell} \right)^{1/2} \exp\left(-\frac{x_\ell^2}{2} \right) \qquad \text{by (5.6),}$$

$$\geqslant \frac{3}{4} \exp\left(-\frac{1}{2} \right) \sum_{\ell \in \Delta_n} c(\ell) \cdot \left(\frac{2}{\pi n_\ell} \right)^{1/2} \qquad \text{by (5.14),}$$

$$\geqslant \frac{3}{4\sqrt{e}} \sum_{\ell \in \Delta_n} c(\ell) \cdot \left(\frac{2-\alpha}{\pi n} \right)^{1/2} \qquad \text{by (5.17),}$$

$$\geqslant c_5 \cdot \min_{\ell \in \Delta_n} c(\ell) \cdot \frac{|\Delta_n|}{\sqrt{n}} \qquad \qquad \text{with } c_5 = \frac{3}{4} \left(\frac{2-\alpha}{\pi e} \right)^{1/2},$$

$$\geqslant c_5 \cdot c_4 \cdot \min_{\ell \in \Delta_n} c(\ell) \qquad \qquad \text{by (5.16).}$$

By (5.7) and (5.15), we see that this last quantity is bounded away from zero for sufficiently large $n \in \mathbb{N}$, which completes the proof. $\square$

Why $3n + 1$ and not $pn + 1$? There is an interesting heuristic argument why the dynamical system on $\mathbb{N}$ generated by the $3n + 1$ function should by special among those generated by a $pn + 1$ function (for an odd prime p)

$$T_p(n) := \begin{cases} n/2 & \text{if } n \text{ is even,} \\ (pn + 1)/2 & \text{if } n \text{ is odd.} \end{cases}$$

Suppose that an averaging procedure involving p-adic averages is adequate for a description of the dynamical behaviour of iterates of T_p (this assumption is, to be precise, not yet proved even for the $3n + 1$ function $T = T_3$). Then we are lead to a sequence of "averaged estimating series"

$$\bar{s}_{n,p} = \sum_{\ell=1}^{\infty} \frac{1}{(p-1)p^{\ell-1}} \binom{n + \lfloor (\log_2 p)\ell \rfloor}{\ell}.$$

(Here the factor $(p-1)p^{\ell-1}$ is just the order of the multiplicative group of prime residues modulo p^ℓ.) Computing the index $\ell_p(n)$ of the maximal term by the obvious generalization of the formula given in theorem 4.3 yields

$$\lim_{n \to \infty} \frac{\ell_p(n)}{n} = \frac{1}{2 - \log_2 p} < 0 \qquad \text{if } p > 4.$$

Thus, a proof like that of theorem 5.2 would be impossible. Presumably, we would be led to a formula like

$$\limsup_{n \to \infty} \frac{\bar{s}_{n,p}}{2^n} = 0,$$

which would mean, by the above assumption, that the weak components of the Collatz graph of T_p corresponding to limiting cycles cannot have positive asymptotic density. And this would have the consequence that, in the dynamical system on $\mathbb{N}$ generated by T_p, there are either divergent trajectories, or infinitely many different limiting cycles.

Not precisely this hypothetical consequence, but the fact that for *Wieferich primes* p, iteration of T_p also admits other limiting behaviour than a T_p-cycle containing 1, is proved by Brocco [**Bro**] (1995).

CHAPTER IV

AN ASYMPTOTICALLY HOMOGENEOUS MARKOV CHAIN

The purpose of this chapter is to show that the counting functions of admissible vectors

$$e_\ell(k, a) = |\mathcal{E}_{\ell,k}(a)|$$

can be constructed using an asymptotically homogeneous discrete-time Markov process. To get the idea behind that, consider the production rule

$$e_{\ell+1}(k, a) = \sum_{j=0}^{k} e_\ell \left(k - j, \frac{2^{j+1}a - 1}{3} \right) .$$

Now the intuitive idea is that the sum in this production rule of the counting functions is, in fact, a hidden integration; the actual construction is technically involved.

In the first section, we introduce *small* admissible vectors. Then the counting functions $e_\ell(k, a)$ of admissible vectors are expressible as a Cauchy product of a sequence of partition functions, and counting functions $g_\ell(k, a)$ of small admissible vectors. The latter counting functions have the advantage of having, in some sense, 'compact support', which makes them apt for a renormalization procedure.

The—technically complicated—construction of the state space of the Markov chain is done in section 2. The state space is written as a product $\mathbb{B}_3 \times \mathbb{Z}_3^*$ of two factors, corresponding to the two variables of the counting functions. The first factor is the set $\mathbb{B}_3$ of 3-*adic fractions*, and the second factor is the group $\mathbb{Z}_3^*$. The state space is intrinsically uncountable. We shall furnish it with a delicately chosen topology, to make the renormalized counterparts of the counting functions continuous. There is also a natural measure on that state space, which we shall use as a reference measure throughout.

At this stage, it is more or less straightforward to construct the sequence of transition probabilities. It is only necessary to bear in mind that we want to view the renormalized versions of the counting functions as successive densities of the Markov chain, w.r.t. the given reference measure. In section 3, we compute a sequence of integral kernels in terms of which the transition probabilities are given.

Up to then, we have constructed a non-homogeneous Markov chain by explicitly describing the sequence of families of transition measures. In section 4, we prove that this Markov chain is asymptotically homogeneous in the sense that, for each point $(x, a) \in \mathbb{B}_3 \times \mathbb{Z}_3^*$, the sequence of transition measures associated to this point converges vaguely to a probability measure. It turns out that each of these limiting probability measures is averaging on the second factor $\mathbb{Z}_3^*$. The family of all limiting pobability measures associated to the points of the state space $\mathbb{B}_3 \times \mathbb{Z}_3^*$ form a *limiting transition probability*.

In section 5, it is shown that this limiting transition probability comes from a transition probability on the real unit interval as state space, having there a uniquely determined invariant density. Moreover, it is shown that this invariant density is a C^∞-function on the real line, which has compact support, and which is a polynomial on each interval outside the classical Cantor set.

In the final section 6, we give some heuristic considerations concerning possible relations between this asymptotically homogeneous Markov chain and $3n + 1$ predecessor densities.

1. Small vectors and a Cauchy product

In this section, we pursue an idea originating in the investigations of section II.3. Remember that there is an equivalence relation on the set of feasible vectors, namely the similarity $s \simeq t$ defined in II.3.6. Remember also that, by lemma II.3.7, this relation respects admissibility in the sense that $s \in \mathcal{E}(a)$ and $s \simeq t$ imply $t \in \mathcal{E}(a)$, for any $a \in \mathbb{N}$. Now it is possible to "generate" a similarity class by taking ist smallest element (i.e. that with minimal absolute value) and adding appropriate non-negative integer vectors to it. This procedure is described here in some detail; it makes it ultimately possible to compute the sequence of counting function $(e_\ell)_{\ell \in \mathbb{N}}$ defined II.4.1 as a Cauchy product of the sequence of counting functions of those smallest elements, and of a sequence of partition functions.

1.1. DEFINITION. A feasible vector $(s_0, \ldots, s_\ell) \in \mathcal{F}$ is called *small vector*, if $s_j < 2 \cdot 3^{j-1}$ for $j = 0, \ldots, \ell$. The subset $\mathcal{S} \subset \mathcal{F}$ is defined to be the set of small vectors. For given $a \in \mathbb{N}$, $k, \ell \in \mathbb{N}_0$, we use the notations

$$\mathcal{G}(a) := \mathcal{E}(a) \cap \mathcal{S},$$
$$\mathcal{G}_\ell(a) := \{s \in \mathcal{G}(a) : \ell(s) = \ell\},$$
$$\mathcal{G}_{\ell,k}(a) := \mathcal{E}_{\ell,k}(a) \cap \mathcal{S} = \{s \in \mathcal{G}(a) : \ell(s) = \ell, |s| = k\}.$$

According to definition II.2.2, a feasible vector consists of non-negative integers, which implies for small vectors $(s_0, \ldots, s_\ell) \in \mathcal{S}$ that

$$(1.1) \qquad s_0 = 0, \quad s_1 \in \{0, 1\}, \quad \ldots, \quad s_\ell \in \{0, \ldots, 2 \cdot 3^{\ell-1} - 1\}.$$

From this we infer immediately

$$(1.2) \qquad |s| \leqslant \sum_{j=1}^{\ell(s)} (2 \cdot 3^{j-1} - 1) = 3^{\ell(s)} - \ell(s) - 1 \qquad \text{for any } s \in \mathcal{S} \setminus \{(0)\}.$$

The structure of similarity classes. Recall that, according to definition II.3.6, two feasible vectors $s, t \in \mathcal{F}$ are called similar, if they have equal length, and if their components satisfy $s_j \equiv t_j \mod 2 \cdot 3^{j-1}$ for $j = 1, \ldots, \ell(s) = \ell(t)$. Here we give a closer analysis of this equivalence relation.

1.2. DEFINITION. Let $\ell \in \mathbb{N}_0$. An integer vector $h = (h_0, \ldots, h_\ell) \in \mathbb{Z}^{\ell+1}$ (not necessarily non-negative) will be called *difference vector of length* ℓ, if $h_j \equiv 0 \mod 2 \cdot 3^{\ell-1}$ for each $j \in \{1, \ldots, \ell\}$ (h_0 is not restricted). We denote the set of all difference vectors of length ℓ by $\mathcal{D}_\ell \subset \mathbb{Z}^{\ell+1}$; and $\mathcal{D}_\ell^+ = \mathcal{D}_\ell \cap \mathcal{F}$ will denote the subset of non-negative difference vectors. In addition, we will use freely the notions *length, absolute value,* and *norm* (introduced in definition II.2.2) also for difference vectors.

1.3. REMARK. Difference vectors arise as differences of similar feasible vectors. Formally, let $s, t \in \mathcal{F}$ with $\ell(s) = \ell(t) =: \ell$. Then their difference $s - t$ is well-defined in $\mathbb{Z}^{\ell+1}$, and

$$s \simeq t \quad \Longleftrightarrow \quad s - t \in \mathcal{D}_\ell .$$

Now the idea is to describe the structure of a similarity class $\{t\}^\simeq = \{u \in \mathcal{F} : u \simeq t\}$ by generating it out of its smallest element.

1.4. LEMMA. *There is a unique projection* $\gamma : \mathcal{F} \to \mathcal{S}$ *satisfying* $\gamma(t) \simeq t$ *for any* $t \in \mathcal{F}$. *Moreover, this projection is given by* $\gamma(t) = t - h(t)$ *where*

$$h_0(t) = t_0, \qquad h_j(t) = 2 \cdot 3^{j-1} \left\lfloor \frac{t_j}{2 \cdot 3^{j-1}} \right\rfloor \qquad \textit{for } j = 1, \ldots, \ell,$$

and γ *has the following properties (for arbitrary* $s, t \in \mathcal{F}$):

(a) γ *is surjective.*
(b) $s \simeq t$ *implies* $\gamma(s) = \gamma(t)$.
(c) $|\gamma(t)| = \min \{|u| : u \in \{t\}^\simeq\}$.
(d) $\gamma^{-1}(s) = \{s\}^\simeq = s + \mathcal{D}_{\ell(s)}^+$ *for any* $s \in \mathcal{S}$.

PROOF. Let $t = (t_0, \ldots, t_\ell) \in \mathcal{F}$. Then $\gamma(t) = t - h(t) = (0, r_1, \ldots, r_\ell)$ where r_j is just the remainder left by dividing t_j by $2 \cdot 3^{j-1}$, for $j = 1, \ldots, \ell$. Hence, $\gamma(t)$ is a small vector, and uniqueness and all the other properties follow straightforward from the definitions, and from remark 1.3. $\square$

Note that parts (c) and (d) of this lemma encode precisely the generation of a similarity class out of its smallest element: according to part (c), $\gamma(t)$ is the smallest element of the similarity class $\{t\}^\simeq$ (both w.r.t. absolute value and w.r.t. norm), and part (d) says that all feasible vectors which are similar to t are obtained by taking the smallest one, $\gamma(t)$, and adding to it a difference vector of length $\ell(t)$.

Partitions. Our next aim is to compute the number of feasible vectors in one similarity class with previously fixed absolute value. This will be achieved using a special partition function.

1.5. DEFINITION. Let $\ell \in \mathbb{N}$, and fix the constants $c_0 = 1$, $c_j = 2 \cdot 3^{j-1}$ for $j \in \mathbb{N}$. Then we define the special partition function

$$p_\ell(k) := p_{c_1, \dots, c_\ell}(k) := \left| \left\{ (k_0, \dots, k_\ell) \in \mathbb{N}_0^{\ell+1} : \sum_{j=0}^{\ell} k_j c_j = k \right\} \right| \qquad \text{for } k \in \mathbb{N}_0.$$

The usefulness of this special partition function in the present situation arises from the fact that it counts non-negative difference vectors of specified length and absolute value:

$$(1.3) \qquad \left| \{ h \in \mathcal{D}_\ell^+ : |h| = k \} \right| = p_\ell(k) \qquad \text{for } k, \ell \in \mathbb{N}_0.$$

1.6. REMARK. By a well-known procedure in number theory, see, for instance, the book of G. H. Hardy and E. M. Wright [**HW**], section 19.3, we obtain the equation

$$F_\ell(z) := \prod_{j=0}^{\ell} \frac{1}{1 - z^{c_j}} = \sum_{k=0}^{\infty} p_\ell(k) \, z^k$$

which represents the $p_\ell(k)$ as Taylor coefficients of a certain holomorphic function F_ℓ. (It is easy to prove convergence of the infinite product

$$\prod_{j=0}^{\infty} \frac{1}{1 - z^{c_j}} = \sum_{k=0}^{\infty} p_\infty(k) \, z^k \,,$$

where $p_\infty(k) = \sup_{\ell \in \mathbb{N}} p_\ell(k)$ for each $k \in \mathbb{N}_0$.) As each F_ℓ is a holomorphic function in the complex unit disc with singularities at the boundary, Hadamard's formula gives

$$\limsup_{k \to \infty} \sqrt[k]{p_\ell(k)} = 1 \qquad \text{uniformly in } \ell \in \mathbb{N},$$

a relation which we do not really use but which is good to bear in mind in the sequel.

1.7. LEMMA. Let $a \in \mathbb{N}$ and $s \in \mathcal{G}(a)$. Then, for each integer $k \geqslant |s|$,

$$\left| \gamma^{-1}(s) \cap \mathcal{E}_{\ell(s),k}(a) \right| = p_{\ell(s)}(k - |s|).$$

PROOF. According to lemmas 1.4 (d) and II.3.7, we have

$$\gamma^{-1}(s) = s + \mathcal{D}_{\ell(s)}^+ = \{s\}^{\simeq} \subset \mathcal{E}(a).$$

This implies (see definition II.4.1)

$$\gamma^{-1}(s) \cap \mathcal{E}_{\ell(s),k}(a) = \left\{ s + h : h \in \mathcal{D}_\ell^+, |h| = k - |s| \right\}.$$

By (1.3), the latter set has precisely $p_{\ell(s)}(k - |s|)$ elements. $\square$

Counting functions for small admissible vectors. In analogy to the counting functions e_ℓ defined in II.4.1, we consider functions counting small vectors which are admissible w.r.t. some positive integer.

1.8. DEFINITION. Fix $\ell \in \mathbb{N}_0$. Then a *counting function for small admissible vectors* is defined by

$$g_\ell : \mathbb{N}_0 \times \mathbb{N} \to \mathbb{N}_0, \qquad g_\ell(k, a) := |\mathcal{G}_{\ell,k}(a)| \ .$$

The following corollary records some immediate properties of these functions.

1.9. COROLLARY. *For $a, b \in \mathbb{N}$ and $k, \ell \in \mathbb{N}_0$, we have*

(a) *If $\ell \geqslant 1$ and $a \equiv 0 \mod 3$, then $g_\ell(k, a) = 0$.*
(b) *If $a \equiv b \mod 3^\ell$, then $g_\ell(k, a) = g_\ell(k, b)$.*
(c) *If $k \geqslant 3^\ell - \ell$, then $g_\ell(k, a) = 0$.*
(d) *For any $a \in \mathbb{N}$, we have: $g_0(k, a) = 1$, if $k = 0$, and $g_0(k, a) = 0$ otherwise.*

PROOF. (a) and (b) are inherited from properties of the counting functions e_ℓ recorded in lemma 4.2. (c) is a consequence of (1.2), and (d) follows from (1.1). $\square$

For later use, we introduce the following notations.

1.10. NOTATION. For $\ell \in \mathbb{N}$, put $K_\ell := \{0, \dots, 3^\ell - 1\}$ and let $A_\ell^* \subset \mathbb{N}$ be a complete system of incongruent prime residues to modulus 3^ℓ. Moreover, denote the cardinalities of these sets by

$$\Gamma_\ell := |K_\ell| = 3^\ell \qquad \Lambda_\ell := |A_\ell^*| = 2 \cdot 3^{\ell-1} \ .$$

In view of corollary 1.9, the set of values $\{g_\ell(k, a) : k \in K_\ell, a \in A_\ell^*\}$ contains all possible values of g_ℓ, if $\ell \geqslant 1$. This means, in case $\ell \geqslant 1$, the product $K_\ell \times A_\ell^*$ is something like a 'natural domain of definition' of g_ℓ. Observe that by corollary 1.9 (c), in order to comprise the support of g_ℓ, it would have sufficed to take $\{0, \dots, 3^\ell - \ell - 1\}$ instead of K_ℓ. The above choice of K_ℓ is not only dictated by convenience, it also permits an elegant possibility to render the functions φ_ℓ (to be constructed in the next section) continuous.

There will be an occasion to use the following 'total integral' of g_ℓ.

1.11. LEMMA. *Let $\ell \in \mathbb{N}$. Then*

$$\sum_{(k,a) \in K_\ell \times A_\ell^*} g_\ell(k, a) = 2^\ell \cdot 3^{\frac{1}{2}\ell(\ell-1)} = \Lambda_1 \dots \Lambda_\ell \ .$$

PROOF. Let $s \in \mathcal{S}$ be a small vector of length ℓ. By lemma II.3.1, there is exactly one residue class $b \pmod{3^\ell}$ such that s is admissible w.r.t. b. Since

FIGURE 4. The function g_3 on $K_3' \times A_3^*$ with $K_3' = \{0, \ldots, 3^3 - 4\} = \{0, \ldots, 23\}$. The second factor is ordered according to what might be called the *lexicographic order of the inverse Hensel code*: $A_3^* = \{(.100)_3, (.101)_3, (.102)_3, (.110)_3, (.111)_3, (.112)_3, \ldots, (.222)_3\} = \{1, 10, 19, 4, 13, 22, 7, 16, 25, 2, 11, 20, 5, 14, 23, 8, 17, 26\}$.

$\ell \geqslant 1$, we infer from lemma II.3.4 that $b \not\equiv 0 \mod 3$, whence there is exactly one $a \in A_\ell^*$ such that

$$b \equiv a \mod 3^\ell.$$

Together with (1.2), this shows that in the sum of the lemma each small vector of length ℓ is counted exactly once. This implies

$$\sum_{k=0}^{3^\ell - \ell - 1} \sum_{a \in A_\ell^*} g_\ell(k, a) = |\{s \in \mathcal{S} : \ell(s) = \ell\}| = \prod_{j=1}^{\ell} (2 \cdot 3^{j-1}) = 2^\ell \cdot 3^{\frac{1}{2}\ell(\ell-1)}. \quad \square$$

It is interesting to compare the following production rule for the functions g_ℓ with that given in lemma II.4.3.

1.12. LEMMA. *Let* $a \in \mathbb{N}$, $k, \ell \in \mathbb{N}_0$, *and put* $g_\ell(j, q) := 0$ *for* $j < 0$ *or* $q \in (\frac{1}{3}\mathbb{Z}) \setminus \mathbb{Z}$. *Then*

$$g_{\ell+1}(k, a) = \sum_{j=0}^{2 \cdot 3^\ell - 1} g_\ell\left(k - j, \frac{2^{j+1}a - 1}{3}\right).$$

PROOF. (Cf. the proof of lemma II.4.3.) Put $\mathcal{G}_{\ell,j}(q) := \emptyset$ for $j < 0$ or $q \in (\frac{1}{3}\mathbb{Z}) \setminus \mathbb{Z}$. Then there is a one-one-correspondence

$$\mathcal{G}_{\ell+1,k}(a) \quad \longleftrightarrow \quad \bigcup_{j=0}^{2 \cdot 3^{\ell}-1} \mathcal{G}_{\ell,k-j}\left(\frac{2^{j+1}a-1}{3}\right),$$

$$(s_0,\ldots,s_\ell,j) \in \mathcal{G}_{\ell+1,k}(a) \quad \longleftrightarrow \quad (s_0,\ldots,s_\ell) \in \mathcal{G}_{\ell,k-j}\left(\frac{2^{j+1}a-1}{3}\right),$$

from which the claim follows by the definition of g_ℓ. Note that $(s_0,\ldots,s_\ell,j) \in \mathcal{G}(a) \subset \mathcal{S}$ implies by (1.1) that $0 \leqslant j \leqslant 2 \cdot 3^{\ell} - 1$, which entails the restriction of the range of the summation index. □

We use this lemma to evaluate g_1.

1.13. COROLLARY. *The values of g_1 are given by*

$$g_1(k,a) = \begin{cases} 1 & \text{if } k = 0 \text{ and } a \equiv 2 \mod 3, \text{ or if } k = 1 \text{ and } a \equiv 1 \mod 3, \\ 0 & \text{otherwise.} \end{cases}$$

PROOF. By lemma 1.12, we have

$$g_1(k,a) = g_0\left(k, \frac{2a-1}{3}\right) + g_0\left(k-1, \frac{4a-1}{3}\right).$$

By the convention of that lemma, and by part (d) of corollary 1.9, we conclude that $g_0(j,q)$ is non-zero, only if both $j = 0$ and q is an integer. This proves the corollary. □

A Cauchy product.   It is possible to compute a counting function e_ℓ as a combination of p_ℓ and g_ℓ which looks like a Cauchy product; in fact, it is a Cauchy product of two, appropriately defined, sequences.

1.14. THEOREM. *Let $a \in \mathbb{N}$ and $k,\ell \in \mathbb{N}_0$. Then*

$$e_\ell(k,a) = \sum_{j=0}^{k} p_\ell(k-j) \, g_\ell(j,a).$$

PROOF. We have disjoint set decompositions

$$\mathcal{E}_{\ell,k}(a) = \bigcup_{s \in \mathcal{G}_\ell(a)} \gamma^{-1}(s) = \bigcup_{j=0}^{k} \left(\bigcup_{s \in \mathcal{G}_{\ell,j}(a)} \gamma^{-1}(s) \right).$$

Now the summation formula is implied by lemma 1.7. □

2. Renormalization

The purpose of the renormalization procedure presented here is to prove that the counting functions g_ℓ for small admissible vectors can be obtained through an appropriate Markov chain. To this end, we construct a sequence of functions

$$(2.1) \qquad \varphi_\ell : K \times A \longrightarrow \mathbb{R} \qquad \text{for } \ell \in \mathbb{N},$$

which, on one hand, make the counting functions g_ℓ easily recoverable, and which, on the other hand, can be interpreted as successive densities generated by a delicately chosen sequence of transition probabilities. Of course, this only can be achieved, if we ensure that there is a common reference probability measure ϱ on the state space $K \times A$ such that $\int \varphi_\ell d\varrho = 1$ for each $\ell \in \mathbb{N}$. Moreover, we shall require that

(i) both K and A are equipped with a topology,
(ii) p is the tensor product of two naturally defined Borel measures,
(iii) each function φ_ℓ, $\ell \in \mathbb{N}$, is continuous.

Construction of the second factor of the state space. The common domain of definition $K \times A$ of the sequence of functions φ_ℓ, which we are to construct, will be given in such a way that it encorporates, in some sense, all the domains of definition of the counting functions $g_\ell(k, a)$ defined in 1.8. W.r.t. the second factor A, corresponding to the second variable a of g_ℓ, there is, in fact, no problem. We just recall lemma 1.12 and corollary 1.13, and imitate the definition of the counting functions e_ℓ on $\mathbb{N}_0 \times \mathbb{Q}_3$ given in section III.2.

2.1. DEFINITION. The *extended counting functions for small admissible vectors*

$$g_\ell : \mathbb{N}_0 \times \mathbb{Q}_3 \to \mathbb{N}_0, \qquad \text{for } \ell \in \mathbb{N},$$

are defined inductively by

$$g_1(k, a) := \begin{cases} 1 & \text{if } k = 0 \text{ and } a \in \mathbb{Z}_3, \ a \equiv 2 \mod 3, \\ 1 & \text{if } k = 1 \text{ and } a \in \mathbb{Z}_3, \ a \equiv 1 \mod 3, \\ 0 & \text{otherwise.} \end{cases}$$

$$g_{\ell+1}(k, a) := \sum_{j=0}^{2 \cdot 3^\ell - 1} g_\ell \left(k - j, \frac{2^{j+1} a - 1}{3} \right).$$

2.2. REMARK. As in remark III.2.2, we derive from lemma 1.12 that the functions $g_\ell : \mathbb{N}_0 \times \mathbb{Q}_3 \to \mathbb{N}_0$ just defined coincide on $\mathbb{N}_0 \times \mathbb{N} \subset \mathbb{N}_0 \times \mathbb{Q}_3$ with the counting functions of small admissible vectors defined in 1.8. Hence, definition 2.1 gives nothing but an extension of the domain of definition.

We also need the analogue of lemma III.2.3.

2.3. LEMMA. *Let $\ell \in \mathbb{N}$. Then:*

(a) *The support of g_ℓ is contained in $\{0, \ldots, 3^\ell - \ell - 1\} \times \mathbb{Z}_3^*$.*

(b) *If $a, b \in \mathbb{Z}_3$ with $a \equiv b \mod 3^\ell$, then $g_\ell(k, a) = g_\ell(k, b)$ for each $k \in \mathbb{N}_0$.*

PROOF. Repeat the arguments given to prove lemma III.2.3. □

With this in mind, a good candidate for the second factor of the state-space $K \times A$ we are looking for is $A := \mathbb{Z}_3^*$. Indeed, this is a very good candidate because $\mathbb{Z}_3^*$ is already a compact topological group (which involves a natural measure for integration), and the functions $g_\ell(k, \cdot) : \mathbb{Z}_3^* \to \mathbb{N}_0$, being constant on residue classes modulo 3^ℓ, are continuous w.r.t. the 3-adic metric.

Construction of the first factor.   The construction of the first factor of $K \times A$ causes much more trouble. We know from part (c) of corollary 1.9 that, for each $\ell \in \mathbb{N}$ and $a \in \mathbb{Z}_3^*$, the support of $g_\ell(\cdot, a)$ is contained in the finite set $\{0, \ldots, 3^\ell - \ell - 1\}$. The first factor K will be constructed subject to the following conditions.

(i) K is equipped with a total order, a topology, and a natural measure.

(ii) For each $\ell \in \mathbb{N}$, there is an order-preserving projection mapping K onto a subset of $\mathbb{N}_0$ containing the support of the $g_\ell(\cdot, a)$.

(iii) For each $\ell \in \mathbb{N}$, the pull-backs of g_ℓ are continuous on $K \times A$.

Of course, these three properties do not determine K uniquely. They only serve as kind of crash barrier, the main point of the construction being the intuitive idea that the projections should distribute K approximately uniformly onto the sets $\{0, \ldots, 3^\ell - \ell - 1\}$ mentioned above.

2.4. DEFINITION. The set of *3-adic fractions* is given by

$$\mathbb{B}_3 := \left\{ x = (x_j)_{j=1}^\infty : x_j \in \{0, 1, 2\} \text{ for each } j \in \mathbb{N} \right\}.$$

This set is considered to be endowed with the *lexicographic order* (denoted by $\leqslant$), with the order-preserving projection

$$\pi : \mathbb{B}_3 \to [0, 1], \qquad \pi(x) := \sum_{j=1}^\infty \frac{x_j}{3^j},$$

with the topology induced by the *inverse 3-adic metric*

$$d_3'(x, y) := 3^{-\min\{j \in \mathbb{N} : x_j \neq y_j\}},$$

and with the natural measure λ_3 given by

$$\lambda_3(C_{k_1, \ldots, k_\ell}) := 3^{-\ell} \qquad \text{where} \quad C_{k_1, \ldots, k_\ell} := \{x \in \mathbb{B}_3 : x_j = k_j \text{ for } j = 1, \ldots, \ell\}.$$

2.5. REMARK. The reason why we prefer to use $\mathbb{B}_3$ instead of the real unit interval $[0, 1]$ is that the topology of $\mathbb{B}_3$ is, in some sense, finer than that of $[0, 1]$. To be precise, let $0 \leqslant k < 3^\ell$ be an integer represented in digits to basis 3 by

$$k = (k_1 \ldots k_\ell)_3 = \sum_{j=1}^\ell k_j \cdot 3^{\ell - j} \qquad \text{with } k_j \in \{0, 1, 2\} \text{ for } j = 1, \ldots, \ell.$$

With the notations

$$(2.2) \qquad k(0), k(2) \in \mathbb{B}_3, \qquad \begin{aligned} k(0)_j &:= k(2)_j := k_j \quad \text{for } 1 \leqslant j \leqslant \ell, \\ k(0)_j &:= 0, \quad k(2)_j := 2 \quad \text{for } j > \ell, \end{aligned}$$

and with the usual interval notation, understood to be w.r.t. the lexicographic order on $\mathbb{B}_3$, we see that the set

$$C_{k_1,\ldots,k_\ell} = \big[k(0), k(2)\big]$$

is both a closed ball of radius $3^{-\ell-1}$ and an open ball of radius $3^{-\ell}$, centered at each of its elements. On the other hand, evaluating π gives

$$\pi(C_{k_1,\ldots,k_\ell}) = \left[\frac{k}{3^\ell}, \frac{k+1}{3^\ell}\right]$$

which is not an open set in $[0,1] \subset \mathbb{R}$ (for $\ell \geqslant 1$). Thus, we arrive at the following conclusion: a step function on $[0,1]$ with step-length $3^{-\ell}$ cannot be continuous on the unit interval, but it may well be considered as a continuous function on the space $\mathbb{B}_3$ of 3-adic fractions.

2.6. REMARK. When we try to transfer the arithmetic of the real unit interval via π to $\mathbb{B}_3$, we run into very bad arithmetic properties. Addition is definable, if we agree to ignore possible overflow. But such a transfered addition is inevitably discontinuous at points of the type $k(0)$ or $k(2)$ as considered in (2.2). Nevertheless, we shall need some rudimentary arithmetic, e.g., for $x = (x_j)_{j=1}^\infty \in \mathbb{B}_3$,

$$y := \frac{x}{3} \quad \text{is defined by the shift} \quad y_1 := 0, \quad y_j := x_{j-1} \text{ for } j \geqslant 2.$$

Analogously,

$$y := \frac{x+2}{3} \quad \text{is defined by} \quad y_1 := 2, \quad y_j := x_{j-1} \text{ for } j \geqslant 2.$$

These constructions will be used in theorem 4.1.

2.7. REMARK. As a set, as a metric space, and even as a measure space, $\mathbb{B}_3$ is quite similar to the space $\mathbb{Z}_3$ of 3-adic integers given in definition III.1.3, the only difference being that the sequences in $\mathbb{B}_3$ start with index $j = 1$ whereas those in $\mathbb{Z}_3$ start with index $j = 0$. The reason why we prefer to use $\mathbb{B}_3$ in the present situation is to emphasize the ordering aspect. The elements of $\mathbb{B}_3$ are just natural representations of real numbers $x \in [0,1]$, and the natural order in $[0,1]$ is reflected by the lexicographic order on $\mathbb{B}_3$.

2.8. DEFINITION. For each $\ell \in \mathbb{N}$, we use the projection

$$\xi_\ell : \mathbb{B}_3 \to K_\ell, \qquad \xi_\ell(x) := \sum_{j=0}^{\ell-1} x_{\ell-j} \cdot 3^j \quad \in \quad \{0, \ldots, 3^\ell - 1\} = K_\ell .$$

2.9. REMARK. Another description of ξ_ℓ, using the floor function and the projection $\pi : \mathbb{B}_3 \to [0, 1]$, is

$$\xi_\ell(x) = \begin{cases} 3^\ell \pi(x) - 1 & \text{if } x_j = 2 \text{ for } j \geqslant \ell, \\ \lfloor 3^\ell \pi(x) \rfloor & \text{otherwise.} \end{cases}$$

The projections ξ_ℓ are clearly order-preserving. Moreover, when K_ℓ is endowed with the discrete topology, the formula for the inverse 3-adic metric given in definition 2.4 shows that the ξ_ℓ are continuous maps.

The state space and the pull-backs. The state space $K \times A = \mathbb{B}_3 \times \mathbb{Z}_3^*$ is, in fact, a categorial inverse limit (in an appropriate category) of the system $\{K_\ell \times A_\ell^* : \ell \in \mathbb{N}\}$ of the "natural domains of definition" of the functions g_ℓ given in notation 1.10. It is worth to look more closely at the representations of the g_ℓ on $\mathbb{B}_3 \times \mathbb{Z}_3^*$.

2.10. DEFINITION. The *pull-backs* of the counting functions g_ℓ are given by the compositions

$$\widetilde{\varphi}_\ell := g_\ell \circ (\xi_\ell \times \mathrm{id}) : \mathbb{B}_3 \times \mathbb{Z}_3^* \to \mathbb{N}_0, \qquad \widetilde{\varphi}_\ell(x, a) := g_\ell\big(\xi_\ell(x), a\big)$$

2.11. LEMMA. *The pull-backs $\widetilde{\varphi}_\ell$ are continuous functions; moreover, they are constant on the sets $\xi_\ell^{-1}(k) \times \{a \mod 3^\ell\}$ for $k \in K_\ell$, $a \in A_\ell^*$.*

PROOF. Observe that the topologies on $\mathbb{B}_3$ and $\mathbb{Z}_3^*$ have the property that the sets

$$\xi_\ell^{-1}(k) \times \{a \mod 3^\ell\}$$

are both open and closed. Hence, continuity follows, if it is proved that $\widetilde{\varphi}_\ell$ is constant on the above sets. But this is clear from the definition and from lemma 2.3 (b), which asserts that, w.r.t. the second variable, g_ℓ is constant on residue classes modulo 3^ℓ. □

2.12. DEFINITION. The *reference measure* ϱ on $\mathbb{B}_3 \times \mathbb{Z}_3^*$ is defined by the tensor product

$$\varrho := \lambda_3 \otimes \nu_3 ,$$

where λ_3 is given in definition 2.4, and ν_3 is the normalized Haar measure on $\mathbb{Z}_3^*$ given in lemma III.1.8.

2.13. REMARK. For later use, let us compute the measure ϱ on the sets of lemma 2.11.

$$\varrho\big(\xi_\ell^{-1}(k) \times \{a \mod 3^\ell\}\big) = \lambda_3\big(\xi_\ell^{-1}(k)\big) \cdot \nu_3\big(\{a \mod 3^\ell\}\big) = \frac{1}{\Gamma_\ell \Lambda_\ell} = \frac{1}{2 \cdot 3^{2\ell - 1}} .$$

This follows immediately from notation 1.10, definition 2.4, and lemma III.1.8.

The normalization factor. What remains to be done for the construction of the functions φ_ℓ of (2.1) with the required properties, is to put $\varphi_\ell := \gamma_\ell \widetilde{\varphi}_\ell$ with appropriate normalization factors $\gamma_\ell \in \mathbb{R}$.

2.14. LEMMA. *Let $\ell \in \mathbb{N}$, and choose*

$$\gamma_\ell := \frac{\Gamma_\ell}{\Lambda_1 \dots \Lambda_{\ell-1}} = 2^{1-\ell} \cdot 3^{-\frac{1}{2}(\ell^2 - 5\ell + 2)} \,.$$

Then the function

$$\varphi_\ell : \mathbb{B}_3 \times \mathbb{Z}_3^* \to \mathbb{R} \,, \qquad \varphi_\ell(x, a) = \gamma_\ell \, g_\ell\big(\xi_\ell(x), a\big)$$

satisfies

$$\int_{\mathbb{B}_3 \times \mathbb{Z}_3^*} \varphi_\ell(x, a) \, d\varrho(x, a) = 1 \,.$$

PROOF. Putting $\varphi_\ell := \gamma_\ell \, \widetilde{\varphi}_\ell$, we have to choose

$$\gamma_\ell^{-1} = \int_{\mathbb{B}_3 \times \mathbb{Z}_3^*} \widetilde{\varphi}_\ell(x, a) \, d\varrho(x, a) \qquad \text{to ensure } \int \varphi_\ell = 1,$$

$$= \int_{\mathbb{B}_3 \times \mathbb{Z}_3^*} g_\ell\big(\xi_\ell(x), a\big) \, d\varrho(x, a) \qquad \text{by definition 2.10,}$$

$$= \sum_{(k,a) \in K_\ell \times A_\ell^*} g_\ell(k, a) \, \varrho\big(\xi_\ell^{-1}(k) \times \{a \mod 3^\ell\}\big) \qquad \text{by lemma 2.11,}$$

$$= \frac{1}{\Gamma_\ell \Lambda_\ell} \sum_{(k,a) \in K_\ell \times A_\ell^*} g_\ell(k, a) \qquad \text{by remark 2.13,}$$

$$= \frac{\Lambda_1 \dots \Lambda_\ell}{\Gamma_\ell \Lambda_\ell} = 2^{\ell-1} \cdot 3^{\frac{1}{2}(\ell^2 - 5\ell + 2)} \qquad \text{by lemma 1.11.} \quad \square$$

3. Transition probabilities

By now, the sequence $(g_\ell)_{\ell \in \mathbb{N}}$ of counting functions for small admissible vectors has been encoded into a sequence of continuous functions

(3.1) $$\varphi_\ell : \mathbb{B}_3 \times \mathbb{Z}_3^* \to \mathbb{R}, \qquad \varphi_\ell(x, a) := \gamma_\ell \, g_\ell\big(\xi_\ell(x), a\big)$$

with integral 1. It is the aim of this section to view the φ_ℓ as successive densities, w.r.t. the (natural) probability measure $\varrho = \lambda_3 \otimes \nu_3$ on $\mathbb{B}_3 \times \mathbb{Z}_3^*$, of an inhomogeneous (but asymptotically homogeneous) Markov chain on $\mathbb{B}_3 \times \mathbb{Z}_3^*$ starting with the measure

$$\varphi_1 \cdot (\lambda_3 \otimes \nu_3) \,.$$

The method to achieve this is to take the production rule of lemma 1.12 for the g_ℓ, and to translate it carefully to the present situation.

Basic notions for Markov chains.   To fix ideas, let us first recall some basic notions concerning transition probabilities arising in discrete-time Markov processes (which are sometimes called *Markov chains*), cf. D. Revuz [**Rev**].

The *state space* is a set E, together with a σ-algebra $\mathcal{A}$ on E. A *transition probability* on $(E, \mathcal{A})$ is a mapping $P : E \times \mathcal{A} \rightarrow [0, 1]$ with the following properties:

(i) for every fixed $s \in E$, the mapping $\mathcal{A} \rightarrow [0, 1]$, $M \mapsto \tau^s(M) := P(s, M)$ is a measure on $(E, \mathcal{A})$,

(ii) for every fixed $M \in \mathcal{A}$, the function $E \rightarrow [0, 1]$, $s \mapsto f_M(s) := P(s, M)$ is measurable with respect to $\mathcal{A}$.

The measures τ^s of (i) here are refered to as *transition measures*. The transition probability P is called *Markovian*, if each transition measure is a probability measure, i.e. if $\tau^s(E) = P(s, E) = 1$ for each $s \in E$. In this context, we exclusively are concerned with Markovian transition probabilities.

A transition probability P associates to every bounded, measurable function $f : E \rightarrow \mathbb{R}$ the function

$$(3.2) \qquad Pf : E \rightarrow \mathbb{R}, \qquad Pf(s) := \tau^s(f) \,;$$

it is not difficult to verify that Pf is also measurable and that $|Pf| \leq B$, if $|f| \leq B$ for some bound $B > 0$.

Dually, a measure μ on $(E, \mathcal{A})$ is transformed by P into

$$(3.3) \qquad \mu P : \mathcal{A} \rightarrow \mathbb{R}, \qquad \mu P(M) = \mu(f_M),$$

where f_M is the function associated to P as in (ii). Again, it is easy to verify that a Markovian transition probability transforms probability measures into probability measures.

An important class of transition probabilities consists of those whose transition measures are given by some density function with respect to some fixed reference measure: let m denote a fixed probability measure on $(E, \mathcal{A})$, and let p denote a non-negative integrable function on $E \times E$. Then p is called an *integral kernel for a transition probability*, if the mapping

$$(3.4) \qquad P : E \times \mathcal{A} \rightarrow [0, 1], \qquad P(s, M) = \int_M p(s, t) \, dm(t)$$

is a transition probability on $(E, \mathcal{A})$.

We rewrite formula (3.3) in terms of p and a density function ϕ. Assume that the transition probability P is defined via an integral kernel p and that the measure μ has density ϕ with respect to the reference measure m. In this case the transformed measure can also be written with a density w.r.t. m:

$$(3.5) \text{ If } \quad \mu = \phi \cdot m, \quad \text{ then } \quad \mu P = \widetilde{\phi} \cdot m \quad \text{ where } \quad \widetilde{\phi}(t) = \int_E \phi(s) p(s, t) \, dm(s) \,.$$

A density function ψ on E satisfying the integral equation

$$(3.6) \qquad \psi(t) = \int_E \psi(s) p(s, t) \, dm(s)$$

is called a *P-invariant* density.

Domains of dependence and domains of transition. In the situation which is our main concern here, the state space is $E := \mathbb{B}_3 \times \mathbb{Z}_3^*$, together with its Borel σ-algebra. The normalized functions φ_ℓ are considered as densities w.r.t. the reference measure $\varrho = \lambda_3 \otimes \nu_3$, where λ_3 is given in definition 2.4, and ν_3 is the normalized Haar measure on $\mathbb{Z}_3^*$. By lemma 2.14, we infer that

$$\mu_\ell := \varphi_\ell \cdot \varrho, \qquad \ell \in \mathbb{N},$$

is a sequence of probability measures. Our aim is to define integral kernels $p_{\ell,\ell+1}(s,t)$ such that their associated transition probabilities $P_{\ell,\ell+1}$ satisfy

$$(3.7) \qquad \mu_{\ell+1} = \mu_\ell P_{\ell,\ell+1} \,.$$

In other words, and in view of formula (3.5), we are to construct non-negative functions

$$p_{\ell,\ell+1} : (\mathbb{B}_3 \times \mathbb{Z}_3^*) \times (\mathbb{B}_3 \times \mathbb{Z}_3^*) \longrightarrow \mathbb{R}$$

with the property

$$(3.8) \qquad \varphi_{\ell+1}(x', a') = \int_{\mathbb{B}_3 \times \mathbb{Z}_3^*} \varphi_\ell(x, a)\, p_{\ell,\ell+1}\big((x, a), (x', a')\big)\, d\varrho(x, a) \,.$$

In order to do this, we employ a careful analysis of the production rule of the counting functions g_ℓ for small admissible vectors (lemma 1.12),

$$g_{\ell+1}(k, a) := \sum_{j=0}^{2 \cdot 3^\ell - 1} g_\ell\left(k - j, \frac{2^{j+1}a - 1}{3}\right) \,.$$

3.1. DEFINITION. Let $\ell \in \mathbb{N}$ be given, recall notation 1.10, and let $\pi_\ell : \mathbb{Z}_3^* \to A_\ell^*$ denote the natural projection mapping $a \in \mathbb{Z}_3^*$ onto the unique element $\pi_\ell(a) \in A_\ell^*$ which is congruent to a modulo 3^ℓ.

The *domain of dependence* $\Delta_{\ell+1}(k', a')$ of a point $(k', a') \in K_{\ell+1} \times A_{\ell+1}^*$ is defined to be the set of all points $(k, a) \in K_\ell \times A_\ell^*$ where the values $g_\ell(k, a)$ are needed to compute $g_{\ell+1}(k', a')$, i.e.,

$$\Delta_{\ell+1}(k', a') :=$$

$$:= \left\{ (k' - j, a) \in K_\ell \times A_\ell^* \,\Big|\, 0 \leqslant j < \Lambda_{\ell+1},\ a = \pi_\ell\left(\frac{2^{j+1}a' - 1}{3}\right) \in A_\ell^* \right\}$$

$$= \left\{ \left(k, \pi_\ell\left(\frac{2^{j+1}a' - 1}{3}\right)\right) \,\middle|\, \begin{matrix} k' - 1 - \Lambda_{\ell+1} \leqslant k \leqslant k',\, 0 \leqslant k \leqslant \Gamma_\ell - 1 \\ 2^{j+1}a' \equiv 1 \mod 3,\, 2^{j+1}a' \not\equiv 1 \mod 9 \end{matrix} \right\} \,.$$

Dually, the *domain of transition* of a point $(k, a) \in K_\ell \times A_\ell^*$ is defined to be the set of all points $(k', a') \in K_{\ell+1} \times A_{\ell+1}^*$ where the values $g_{\ell+1}(k', a')$ are influenced by $g_\ell(k, a)$, i.e.,

$$(3.9)$$

$$\Delta_\ell^*(k, a) := \left\{ (k', a') \in K_{\ell+1} \times A_{\ell+1}^* : (k, a) \in \Delta_{\ell+1}(k', a') \right\}$$

$$= \left\{ (k + j, a') : 0 \leqslant j < \Lambda_{\ell+1},\, 2^{j+1}a' \equiv 3a + 1 \mod 3^{\ell+1} \right\} \,.$$

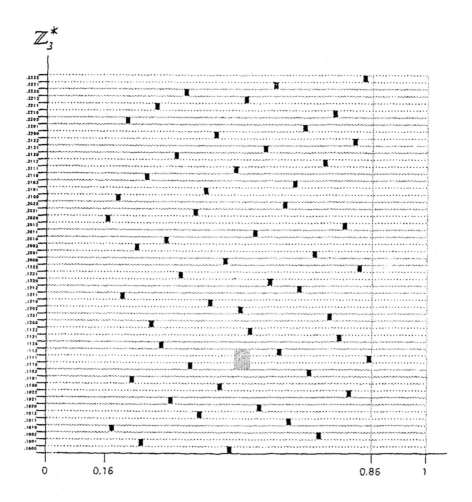

FIGURE 5. The area consisting of the black rectangles shows the domain of transition $\widetilde{\Delta}_3^*(x, a)$ for $x = \pi^{-1}\left(\frac{1}{2}\right) = (1, 1, \dots) \in \mathbb{B}_3$ corresponding to $k = \xi_3(x) = 13$, and $a = (1, 1, \dots) \in \mathbb{Z}_3^*$; the gray area shows $\xi_3^{-1}(k) \times \pi_3^{-1}(a)$. The labels on the a-axis are given in the *inverse Hensel code*, e.g. .1211 corresponds to $43 = 1 + 2\cdot3 + 1\cdot3^2 + 1\cdot3^3$. The labels on the x-axis denote $\pi(x) \in [0, 1]$.

The *lifted domain of dependence* and the *lifted domain of transition* are defined to be the inverse images of the domains in the state space $\mathbb{B}_3 \times \mathbb{Z}_3^*$,

$$\widetilde{\Delta}_{\ell+1}(x', a') := (\xi_\ell \times \pi_\ell)^{-1} \left(\Delta_{\ell+1}(\xi_{\ell+1}(x'), \pi_{\ell+1}(a')) \right)$$

$$= \bigcup \left\{ \xi_\ell^{-1}(k) \times \pi_\ell^{-1}(a) : (k,a) \in \Delta_{\ell+1}(\xi_{\ell+1}(x'), \pi_{\ell+1}(a')) \right\} ,$$

$$\widetilde{\Delta}_\ell^*(x, a) := (\xi_{\ell+1} \times \pi_{\ell+1})^{-1} \left(\Delta_\ell^*(\xi_\ell(x), \pi_\ell(a)) \right)$$

$$= \bigcup \left\{ \xi_{\ell+1}^{-1}(k') \times \pi_{\ell+1}^{-1}(a') : (k', a') \in \Delta_\ell^*(\xi_\ell(k), \pi_\ell(a)) \right\} .$$

3.2. REMARK. Of course, there is a set equation for the lifted domains of transition which is analogous to (3.9):

$$\widetilde{\Delta}_\ell^*(x, a) = \left\{ (x', a') \in \mathbb{B}_3 \times \mathbb{Z}_3^* : (x, a) \in \widetilde{\Delta}_{\ell+1}(x', a') \right\} .$$

The integral kernels. We now reformulate the production rule for the counting functions g_ℓ for small admissible vectors, given in lemma 1.12, in terms of the domain of dependence

$$(3.10) \qquad g_{\ell+1}(k', a') = \sum_{(k,a) \in \Delta_{\ell+1}(k', a')} g_\ell(k, a) .$$

Our next task is to translate this into the language of transition probabilities.

3.3. LEMMA. *The sequence of transition probabilities* $P_{\ell, \ell+1}$ *defined by the integral kernels*

$$p_{\ell, \ell+1}\big((x, a), (x', a')\big) := \begin{cases} \Gamma_{\ell+1} & \text{if } (x', a') \in \widetilde{\Delta}_\ell^*(x, a), \\ 0 & \text{otherwise,} \end{cases}$$

satisfies the relation $\mu_{\ell+1} = \mu_\ell P_{\ell, \ell+1}$ *with* $\mu_\ell = \varphi_\ell \cdot \varrho$.

PROOF. Let $(x', a') \in \mathbb{B}_3 \times \mathbb{Z}_3^*$, and put for abbreviation $k' = \xi_{\ell+1}(x')$ and $a' = \pi_{\ell+1}(a')$ (there is no risk of confusion). By (3.10) and the definition of the lifted domain of dependence, we derive

$$\varphi_{\ell+1}(x', a') = \gamma_{\ell+1} \, g_{\ell+1}(k', a') = \gamma_{\ell+1} \sum_{(k,a) \in \Delta_{\ell+1}(k', a')} g_\ell(k, a)$$

$$(3.11) \qquad\qquad = \frac{\gamma_{\ell+1}}{\gamma_\ell} \sum_{(k,a) \in \Delta_{\ell+1}(k', a')} \varphi_\ell\big(\xi_\ell^{-1}(k), \pi_\ell^{-1}(a)\big) .$$

This notation makes sense, as, by lemma 2.11, the function $\varphi_\ell = \widetilde{\varphi}_\ell$ is constant on each set $\xi_\ell^{-1}(k) \times \pi_\ell^{-1}(a) \subset \mathbb{B}_3 \times \mathbb{Z}_3^*$.

To write the sum as an integral we make use of the formula for the reference measure ϱ given in remark 2.13. We continue the calculation started in (3.11):

$$\varphi_{\ell+1}(x', a') = \ldots = \frac{\gamma_{\ell+1} \Gamma_\ell \Lambda_\ell}{\gamma_\ell} \int_{\widetilde{\Delta}_{\ell+1}(x', a')} \varphi_\ell(x, a) \, d\varrho(x, a) .$$

On the other hand, by (3.8), the integral kernel $p_{\ell,\ell+1}$ for which we are looking is to fulfill

$$\varphi_{\ell+1}(x', a') = \int_{\mathbb{B}_3 \times \mathbb{Z}_3^*} \varphi_\ell(x, a) p_{\ell,\ell+1}\big((x, a), (x', a')\big) \, d\varrho(x, a) \,.$$

This means that $p_{\ell,\ell+1}$ should have the following property: If we fix $(x', a') \in \mathbb{B}_3 \times \mathbb{Z}_3^*$, then the function

$$(x, a) \mapsto p_{\ell,\ell+1}\big((x, a), (x', a')\big)$$

should have support $\widetilde{\Delta}_{\ell+1}(x', a')$, assuming there the constant value

$$\frac{\gamma_{\ell+1}\Gamma_\ell\Lambda_\ell}{\gamma_\ell} = \frac{\Gamma_{\ell+1}}{\Lambda_1 \ldots \Lambda_\ell} \cdot \frac{\Lambda_1 \ldots \Lambda_{\ell-1}}{\Gamma_\ell} \cdot \Gamma_\ell\Lambda_\ell = \Gamma_{\ell+1} \,.$$

Writing $(x', a') \in \widetilde{\Delta}_\ell^*(x, a)$ instead of $(x, a) \in \widetilde{\Delta}_{\ell+1}(x', a')$, the lemma follows. $\square$

3.4. COROLLARY. *The sequence of density functions φ_ℓ satisfies*

$$\varphi_{\ell+1}(x', a') = \int_{\widetilde{\Delta}_{\ell+1}(x', a')} \Gamma_{\ell+1}\varphi_\ell(x, a) \, d\varrho(x, a) \,.$$

4. Vague convergence of the transition measures

In the preceding section we have defined a sequence of transition probabilities $P_{\ell,\ell+1}$ on the state space $\mathbb{B}_3 \times \mathbb{Z}_3^*$ (for notational convenience, we assume $\ell \geqslant 2$ and consider transitions $\ell-1 \to \ell$ in this section). According to (3.4) and lemma 3.3, the associated transition measures can be computed to

(4.1) $$\tau_{\ell-1,\ell}^{(x,a)} = \Gamma_\ell \, \varrho|_{\widetilde{\Delta}_{\ell-1}^*(x,a)} \,, \qquad \text{for } \ell \geqslant 2 \,.$$

We shall prove here that these transition measures converge vaguely to a family of transition measures forming a "limiting" transition probability on $\mathbb{B}_3 \times \mathbb{Z}_3^*$. It will also be seen that this vague convergence is uniform w.r.t. $(x, a) \in \mathbb{B}_3 \times \mathbb{Z}_3^*$, and, moreover, that it is uniform on bounded equicontinuous families of functions $f : \mathbb{B}_3 \times \mathbb{Z}_3^* \to \mathbb{R}$.

Let us recall precise definitions. A sequence $(\mu_n)_{n \in \mathbb{N}}$ of (Borel) measures on a topological space X is said to converge *vaguely* to a measure μ, if, for every continuous function $f : X \to \mathbb{R}$, we have $\lim_{n \to \infty} \mu_n(f) = \mu(f)$.

To deal with equicontinuity, we use the *(global) modulus of continuity* of a continuous function f on a metric space (X, d):

$$\omega_f(\delta) := \sup\{|f(x) - f(y)| : d(x, y) < \delta\} \qquad \text{for } \delta > 0 \,.$$

A family $\mathcal{F}$ of continuous functions $f : X \to \mathbb{R}$ is called *(globally) equicontinuous*, if

$$\lim_{\delta \to 0} \left(\sup_{f \in \mathcal{F}} \omega_f(\delta) \right) = 0 \,.$$

In the present situation, $\mathbb{B}_3 \times \mathbb{Z}_3^*$ is a compact metric space. We take the metric to be defined by

$$(4.2) \qquad d\big((x, a), (y, b)\big) := \max\{d_3'(x, y), d_3(a, b)\}$$

with the metric d_3' on the set $\mathbb{B}_3$ of 3-adic fractions as given in definition 2.4, and with the usual 3-adic metric d_3 on $\mathbb{Z}_3^*$, see lemma & definition III.1.2. The vague limit of the transition measures will be expressed in terms of the measure λ_3 on $\mathbb{B}_3$, as defined in 2.4, and of the Haar measure ν_3 on $\mathbb{Z}_3^*$. We shall also use the interval notation w.r.t. the lexicographic order on $\mathbb{B}_3$, and the rudiments of arithmetic in $\mathbb{B}_3$ described in remark 2.6.

4.1. THEOREM. *For every $(x, a) \in \mathbb{B}_3 \times \mathbb{Z}_3^*$, the sequence of transition measures converges vaguely,*

$$\tau_{\ell-1,\ell}^{(x,a)} \to \tau^x := \left. \frac{3}{2} \lambda_3 \right|_{\left[\frac{x}{3}, \frac{x+2}{3} \right]} \otimes \nu_3 \,, \qquad as \quad \ell \to \infty \,.$$

This convergence is uniform w.r.t. $(x, a) \in \mathbb{B}_3 \times \mathbb{Z}_3^*$. Moreover, it is uniform on any bounded equicontinuous family of functions $f : \mathbb{B}_3 \times \mathbb{Z}_3^* \to \mathbb{R}$.

PROOF. Let $\mathcal{F}$ be a bounded equicontinuous family of real functions on $\mathbb{B}_3 \times \mathbb{Z}_3^*$. This means that there is a constant $B > 0$ such that $|f| \leqslant B$ for each $f \in \mathcal{F}$ and that the function

$$(4.3) \qquad s(\delta) := \sup_{f \in \mathcal{F}} \omega_f(\delta) \qquad \text{satisfies} \qquad \lim_{\delta \to 0} s(\delta) = 0 \,.$$

We are going to show that, for any given $\varepsilon > 0$, there is a number $N(\varepsilon) \in \mathbb{N}$ such that

$$(4.4) \qquad \left| \tau_{\ell-1,\ell}^{(x,a)}(f) - \tau^x(f) \right| < \varepsilon \qquad \text{for} \quad \ell > N(\varepsilon), \ (x, a) \in \mathbb{B}_3 \times \mathbb{Z}_3^*, \ f \in \mathcal{F} \,.$$

Before diving into the estimates, let us give some technical prerequisites designed to trace out the domains of dependence $\widetilde{\Delta}_{\ell-1}^*(x, a) \subset \mathbb{B}_3 \times \mathbb{Z}_3^*$. We need an auxiliary (large) $N_0 \in \mathbb{N}$ which will be fixed later in dependence of ε. An essential role in the proof will be played by the intervals (in $\mathbb{B}_3$)

$$(4.5) \qquad J_\ell(k, m) := \bigcup_{0 \leq i < \Lambda_{N_0}} \xi_\ell^{-1}\left(k + m\Lambda_{N_0} + i\right) \,.$$

Note that $J_\ell(k, m) \subset \mathbb{B}_3$ is well-defined provided

$$(4.6) \qquad \ell > N_0, \quad k \in K_{\ell-1}, \quad 0 \leq m < 3^{\ell-N_0},$$

because we then have the inequalities (the Γ's and Λ's being defined in notation 1.10)

$$0 \leq k + m\Lambda_{N_0} + j < \Gamma_{\ell-1} + \left(3^{\ell-N_0} - 1\right)\Lambda_{N_0} + (\Lambda_{N_0} - 1) < \Gamma_\ell.$$

These intervals come in through the boxes

$$B_m(k, c) := J_\ell(k, m) \times \eta_{N_0}^{-1}(c) \qquad \text{for} \quad c \in A_{N_0}^*,$$

with ℓ, k, m as in (4.6), and with $A_{N_0}^*$ as in notation 1.10.

According to definition 3.1, the domain of transition $\widetilde{\Delta}_{\ell-1}^*(x, a)$ consists of smaller boxes of the type

$$(4.7) \qquad Q_\ell(j, b) := \xi_\ell^{-1}(j) \times \pi_\ell^{-1}(b) \qquad \text{for} \quad j \in K_\ell, b \in A_\ell^*.$$

By the definition of the projections ξ_ℓ in definition 2.8, and the projections π_ℓ in definition 3.1, it is easy to see that

$$\text{either} \quad Q_\ell(j, b) \cap B_m(k, c) = \varnothing \quad \text{or} \quad Q_\ell(j, b) \subset B_m(k, c).$$

The central idea of the proof is contained in the following result:

4.2. LEMMA. *Suppose ℓ, k, m are given as in (4.6), and fix $a \in A_{\ell-1}^*$, $c \in A_{N_0}^*$. Then there is one and only one pair $(j, b) \in \Delta_{\ell-1}^*(k, a)$ such that $Q_\ell(j, b) \subset B_m(k, c)$.*

PROOF. By (4.5) we know that

$$(4.8) \qquad \xi_\ell^{-1}(j) \subset J_\ell(k, m) \iff k + m\Lambda_{N_0} \leq j < k + (m+1)\Lambda_{N_0}.$$

It is well-known from elementary number theory that 2 is a primitive root modulo 3^{N_0}, i.e. that 2 generates the multiplicative group $(\mathbb{Z}/3^{N_0}\mathbb{Z})^*$ of prime residue classes modulo 3^{N_0}. Since $|(\mathbb{Z}/3^{N_0}\mathbb{Z})^*| = \Lambda_{N_0}$, we infer that there is exactly one j satisfying both (4.8) and

$$(4.9) \qquad 2^{j-k+1}c \equiv 3a + 1 \mod 3^{N_0}.$$

Now 2^{j-k+1} is also invertible modulo 3^ℓ, whence one concludes that there is exactly one $b \in A_\ell^*$ such that

$$(4.10) \qquad 2^{j-k+1}b \equiv 3a + 1 \mod 3^\ell.$$

Since (4.9) and (4.10) clearly imply $b \equiv c \mod 3^{N_0}$, we have

$$Q_\ell(j, b) = \xi_\ell^{-1}(j) \times \pi_\ell^{-1}(b) \subset J_\ell(k, m) \times \pi_{N_0}^{-1}(c) = B_m(k, c).$$

By (3.9), this completes the proof of the lemma. $\square$

We shall prove (4.4) by inserting further auxiliary measures

$$\mu_\ell^k := \left.\frac{3}{2}\lambda_3\right|_{J_\ell(k)} \qquad \text{for } \ell \in \mathbb{N},\ k \in K_{\ell-1},$$

where the interval to which λ_3 is restricted is just the union

$$(4.11) \qquad J_\ell(k) := \bigcup_{0 \le m < 3^{\ell-N_0}} J_\ell(k,m) = \bigcup_{0 \le j < \Lambda_\ell} \xi_\ell^{-1}(k+j).$$

This measure is easily evaluated on our boxes:

$$(4.12) \qquad \mu_\ell^k\left(B_m(k,c)\right) = \frac{3}{2}\frac{\Lambda_{N_0}}{\Gamma_\ell}\frac{1}{\Lambda_{N_0}} = \frac{3}{2\cdot\Gamma_\ell} = \Lambda_\ell^{-1}.$$

We can also evaluate the transition measures on these boxes: put $k = \xi_{\ell-1}(x)$, then

$$
\begin{aligned}
\tau_{\ell-1,\ell}^{(x,a)}\left(B_m(k,c)\right) &= \Gamma_\ell\,\varrho\left(\tilde{\Delta}_{\ell-1}^*(x,a)\cap B_m(k,c)\right) && \text{by (4.1),}\\
&= \Gamma_\ell\,\varrho\left(Q_\ell(j,b)\right) && \text{by lemma 4.2,}\\
&= \Gamma_\ell\,\varrho\left(\xi_\ell^{-1}(j)\times\pi_\ell^{-1}(b)\right) && \text{by (4.7),}\\
&(4.13) \qquad = \frac{\Gamma_\ell}{\Gamma_\ell\Lambda_\ell} = \Lambda_\ell^{-1} && \text{by remark 2.13.}
\end{aligned}
$$

We now derive the estimates. Let f_{box} be a bounded, measurable function on $\mathbb{B}_3 \times \mathbb{Z}_3^*$ whose support is contained in a box $B_m(k,c)$. Then (4.12) and (4.13) yield the inequalities

$$
\begin{aligned}
\Lambda_\ell^{-1}\inf f_{\text{box}} &\le \tau_{\ell-1,\ell}^{(x,a)}(f_{\text{box}}) \le \Lambda_\ell^{-1}\sup f_B,\\
\Lambda_\ell^{-1}\inf f_{\text{box}} &\le \mu_\ell^k(f_{\text{box}}) \quad\ \le \Lambda_\ell^{-1}\sup f_B.
\end{aligned}
$$

These imply the estimate on a single box:

$$(4.14) \qquad \left|\tau_{\ell-1,\ell}^{(x,a)}(f_{\text{box}}) - \mu_\ell^k(f_{\text{box}})\right| \le (\sup f_{\text{box}} - \inf f_{\text{box}})\,\Lambda_\ell^{-1}.$$

Let δ be larger than the diameter of a box $B_m(k,c)$ in the metric of (4.2),

$$(4.15) \qquad \delta > \max\left\{\Lambda_{N_0}\Gamma_\ell^{-1}, \Lambda_{N_0}^{-1}\right\} = \max\left\{2\cdot 3^{N_0-1-\ell}, 2^{-1}\cdot 3^{1-N_0}\right\}.$$

Then (4.14) implies for functions $f \in \mathcal{F}$:

$$(4.16) \qquad \left|\left(\tau_{\ell-1,\ell}^{(x,a)} - \mu_\ell^k\right)(f\,\chi_{B_m(k,c)})\right| \le s(\delta)\,\Lambda_\ell^{-1}.$$

To get a global estimate, observe that both measures $\tau_{\ell-1,\ell}^{(x,a)}$ and μ_ℓ^k have support contained in $J_\ell(k) \times \mathbb{Z}_3^*$, where the interval $J_\ell(k)$ is given by (4.11). Furthermore,

$$J_\ell(k) \times A = \bigcup_{0 \leq m < 3^{\ell-N_0}, c \in A_{N_0}^*} B_m(k,c)$$

is a disjoint union of $3^{\ell-N_0}\Lambda_{N_0} = \Lambda_\ell$ boxes. Thus, the following estimate is easily obtained from (4.16) by throwing away $\chi_{B_m(k,c)}$ on the left hand side and multiplying the right hand side by Λ_ℓ:

$$(4.17) \qquad \left| \tau_{\ell-1,\ell}^{(x,a)}(f) - \mu_\ell^k(f) \right| \leq s(\delta).$$

This estimate is true for any $(x,a) \in \mathbb{B}_3 \times \mathbb{Z}_3^*$ and $f \in \mathcal{F}$, provided δ fulfills condition (4.15).

We proceed by estimating the difference between the measures μ_ℓ^k and τ_x. Let

$$I(x) := \left[\frac{x}{3}, \frac{x+2}{3} \right] \subset \mathbb{B}_3$$

denote the interval in $\mathbb{B}_3$ on which the first factor λ_3 of τ^x is restricted. Then the difference between μ_ℓ^k and τ^x on functions $f \in \mathcal{F}$ can be estimated by

$$(4.18) \qquad \left| \mu_\ell^k(f) - \tau^x(f) \right| \leq \frac{3B}{2} \lambda_3 \left(J_\ell(k) \triangle I(x) \right).$$

using the symmetric difference for sets, $X \triangle Y = (X \cup Y) \setminus (X \cap Y)$. But

$$\lambda \left(J_\ell(k) \triangle I(x) \right) = \left| \frac{x}{3} - \frac{[\Gamma_{\ell-1}x]}{\Gamma_\ell} \right| + \left| \frac{x+2}{3} - \frac{[\Gamma_{\ell-1}x] + \Lambda_\ell}{\Gamma_\ell} \right|$$

which clearly tends to 0 if $\ell \to \infty$.

If we combine (4.18) and (4.17) to achieve (4.4), the only remaining problem is the term $s(\delta)$. This can be treated as follows: Using the equicontinuity of $\mathcal{F}$ encoded in (4.3), we conclude that

$$\delta := \sup \left\{ \rho : s(\rho) < \frac{\varepsilon}{2} \right\} > 0.$$

Let us now fix N_0 large enough such that $\delta > \Lambda_{N_0}^{-1}$. Then it is possible to choose $N(\varepsilon)$ large enough to ensure that both (4.15) is satisfied and

$$\frac{3B}{2} \lambda_3 \left(J_\ell(k) \triangle I(x) \right) < \frac{\varepsilon}{2}$$

whenever $\ell > N(\varepsilon)$. This completes the proof of the theorem. $\quad \square$

5. The limiting transition probability

Up to now, the counting functions $g_\ell(k, a)$ for small admissible vectors have been renormalized to yield a sequence of continuous functions

$$\varphi_\ell : \mathbb{B}_3 \times \mathbb{Z}_3^* \to \mathbb{R} \qquad \text{for } \ell \in \mathbb{N},$$

satisfying $\int \varphi_\ell \, d\varrho = 1$. What about the limting behaviour of this sequence, as $\ell \to \infty$?

The invariant density. We already know that the functions φ_ℓ emerge as successive densities in a discrete-time Markov process. Indeed, we saw in corollary 3.4 that

$$\varphi_\ell(x', a') = \int_{\tilde{\Delta}_\ell(x', a')} \Gamma_\ell \, \varphi_{\ell-1}(x, a) \, d\varrho(x, a), \qquad \text{for } \ell \geqslant 2,$$

with the constant $\Gamma_\ell = 3^\ell$. Moreover, it turned out that this Markov chain is asymptotically homogeneous, in the sense that the family of transition measures,

$$\left\{ \tau_{\ell-1, \ell}^{(x, a)} : (x, a) \in \mathbb{B}_3 \times \mathbb{Z}_3^* \right\},$$

converges in some explicitly describable way (see theorem 4.1) to a family of probability measures

$$(5.1) \qquad \left\{ \tau^x = \frac{3}{2} \lambda_3 \Big|_{\left[\frac{x}{3}, \frac{x+2}{3}\right]} \otimes \nu_3 : (x, a) \in \mathbb{B}_3 \times \mathbb{Z}_3^* \right\}.$$

5.1. THEOREM. *The transition probability given by the family of transition measures* (5.1) *has a unique invariant density* $\varphi \in L^1(\mathbb{B}_3 \times \mathbb{Z}_3^*, \varrho)$ *given by*

$$\varphi(x, a) = \psi(\pi(x)) \qquad \varrho\text{-a.e.,}$$

where the function $\psi \in L^1(\mathbb{R})$ *is uniquely determined by the following conditions:*

(i) $\qquad\qquad \text{supp}(\psi) \subset [0, 1].$

(ii) $\qquad\qquad \displaystyle\int_0^1 \psi(s) \, ds = 1.$

(iii) $\qquad\qquad \displaystyle\psi(t) = \frac{3}{2} \int_{3t-2}^{3t} \psi(s) \, ds \qquad \text{for any } t \in [0, 1].$

PROOF. The limiting transition measures τ^x are averaging w.r.t. the second variable; they do not depend on $a \in \mathbb{Z}_3^*$. As the projection $\pi : \mathbb{B}_3 \to [0, 1]$ of definition 2.4 is measure-preserving, we may think of τ^x as being given by a family of transition measures

$$\mathcal{M} := \{ \mu^s : s \in [0, 1] \}$$

on the real unit interval, the μ^s being defined by

$$\mu^{\pi(x)}(\pi \circ f) := \tau^x(f)$$

for each continuous function $f : \mathbb{B}_3 \to \mathbb{R}$. This definition is unambiguous outside a set of measure 0. It defines μ^s on step functions on $[0, 1]$ with step-length $3^{-\ell}$ (for some $\ell \in \mathbb{N}$; see remark 2.5), which suffices to fix μ^s as a Borel measure.

The formula for τ^x given in theorem 4.1 implies that each μ^s is given by the density

$$\frac{3}{2} \chi_{\left[\frac{s}{3}, \frac{s+2}{3}\right]}$$

w.r.t. the Lebesgue measure λ on $[0, 1]$. We obtain that the transition probability P defined by the family $\mathcal{M}$ of transition measures is given by the integral kernel

$$p(s, t) := \frac{3}{2} \chi_{\left[\frac{s}{3}, \frac{s+2}{3}\right]}(t) = \begin{cases} \frac{3}{2} & \text{if } t \in [0, 1] \text{ and } 3t - 2 \leqslant s \leqslant 3t, \\ 0 & \text{otherwise,} \end{cases}$$

for $s, t \in [0, 1]$. By formula (3.6), we conclude that an invariant density of the transition probability P must be a fixed point of the operator

$$P^* : L^1([0, 1]) \to L^1([0, 1]), \qquad P^* f(t) := \frac{3}{2} \int_{3t-2}^{3t} f(s)\, ds,$$

where, in order to evaluate the integral, each $f \in L^1([0, 1])$ is supposed to be extended by 0 outside $[0, 1]$. (It is not difficult to verify by a short calculation that P^* is integral-preserving; hence it is indeed a Markov operator on $L^1([0, 1])$.)

Next we show that P^* is norm-contractive on the hyperplane of $L^1([0, 1])$ consisting of the functions with vanishing integral. Let $f \in L^1([0, 1])$ with $\int_0^1 f(t) dt = 0$. Then

$$\|P^* f\|_1 =$$

$$= \int_0^{\frac{1}{3}} \left| \frac{3}{2} \int_0^{3t} f(s)\, ds \right| dt + \int_{\frac{1}{3}}^{\frac{2}{3}} \left| \frac{3}{2} \int_0^1 f(s)\, ds \right| dt + \int_{\frac{2}{3}}^1 \left| \frac{3}{2} \int_{3t-2}^1 f(s)\, ds \right| dt$$

$$= \frac{3}{2} \int_0^{\frac{1}{3}} \left(\left| \int_0^{3t} f(s)\, ds \right| + \left| \int_{3t}^1 f(s)\, ds \right| \right) dt$$

$$\leqslant \frac{1}{2} \|f\|_1 .$$

We conclude, using Banach's fixed point theorem, that there is a unique $\psi \in L^1([0, 1])$ with $\int \psi(s) ds = 1$ satisfying $P^* \psi = \psi$. This completes the proof. $\square$

5.2. REMARK. This proof shows that the function ψ determined by conditions (i)—(iii) of the theorem is non-negative. Indeed, the operator P^* maps non-negative functions onto non-negative ones. As P^* is a contraction on a hyperplane, the invariant density ψ may be constructed as the limit of a sequence $(\psi_k)_{k \in \mathbb{N}_0}$ starting with an arbitrary initial function $\psi_0 \in L^1([0, 1])$ with $\int_0^1 \psi_0(x)\, dx = 1$. Taking $\psi_0 \equiv 1$, we obtain a sequence of non-negative L^1-functions converging to ψ, whence $\psi \geqslant 0$.

A relation to Cantor's set. The function ψ determined by conditions (i)—(iii) of the theorem has a nice relation to the classical 'Cantor middle thirds set,' which we are now going to describe. For our purpose, an appropriate starting point is to consider the dilations (cf. [**Edg**])

$$f_1(x) := \frac{x}{3}, \qquad f_2(x) := \frac{x+2}{3}.$$

Now put

(5.2) $\qquad C_0 := [0, 1], \qquad C_n := f_1(C_{n-1}) \cup f_2(C_{n-1}) \quad \text{for} \quad n \geq 1.$

In other words: to obtain C_1, we have to remove the middle third

$$M := \left\{ \frac{1}{3} < x < \frac{2}{3} \right\}$$

from the unit interval C_0. Clearly, M is an interval of length $\delta := 1/3$, centered about the center of $[0, 1]$. To construct C_n out of C_{n-1} according to (5.2), we have to squeeze each interval of C_{n-1} using f_1 and f_2 and put it once on the left and once on the right of M. The effect is the same as if we had removed an interval of proportional length δ from the middle of each interval of C_{n-1}.

5.3. LEMMA. *Let $\psi : \mathbb{R} \to \mathbb{R}$ be the function defined by conditions (i)—(iii) of theorem 5.1. Then ψ is a $\mathcal{C}^\infty$-function with compact support. Moreover, for each $n \in \mathbb{N}_0$, C_n is the support of the n-th derivative of ψ.*

PROOF. First note that ψ is a $\mathcal{C}^\infty$-function because it is a solution of the integral equation (iii). Let Ψ be a primitive of ψ. Then condition (iii) implies

$$\psi(t) = \frac{3}{2} \left(\Psi(3t) - \Psi(3t - 2) \right),$$

and differentiation gives

(5.3) $\qquad\qquad \psi'(t) = \frac{3^2}{2} \left(\psi(3t) - \psi(3t - 2) \right).$

Denoting the interior of C_n by $\text{int}\, C_n$, we now prove by induction on n that $\psi^{(n)}(t) \neq 0$ for any $t \in \text{int}\, C_n$.

$n = 0$: We know by remark 5.2 that $\psi \geq 0$. By the normalization condition (ii) we conclude that there is a point $t_0 \in]0, 1[$ such that $\psi(t_0) > 0$. Then condition (iii) proves that $\psi(t) > 0$ whenever

$$\frac{t_0}{3} < t < \frac{t_0 + 2}{3}.$$

Thus we finally arrive, inductively, at the conclusion that $\psi(t) > 0$ whenever $0 < t < 1$.

$n - 1 \to n$: Differentiating (5.3) $(n - 1)$ times, we get

$$(5.4) \qquad \psi^{(n)}(x) = \frac{3^{n+1}}{2} \left(\psi^{(n-1)}(3x) - \psi^{(n-1)}(3x - 2) \right).$$

By the induction hypothesis, we know that supp $\left(\psi^{(n-1)} \right) = C_{n-1} \subset [0, 1]$. This implies that there is at most one non-zero term in the difference in (5.4). Again by the induction hypothesis, we have

$$\psi^{(n-1)}(3x) \neq 0 \quad \Longleftrightarrow \quad 3x \in \text{int } C_{n-1} \quad \Longleftrightarrow \quad x \in f_1(\text{int } C_{n-1}),$$
$$\psi^{(n-1)}(3x - 2) \neq 0 \quad \Longleftrightarrow \quad 3x - 2 \in \text{int } C_{n-1} \quad \Longleftrightarrow \quad x \in f_2(\text{int } C_{n-1}).$$

By (5.2), this completes the proof. $\square$

The function ψ of lemma 5.3 is really an astonishing one: it is a C^∞-function with compact support, which is 'piecewise a polynomial.' More precisely: if we define a sequence $(V_n)_{n \geqslant 0}$ of open subsets of $[0, 1]$ inductively by

$$V_0 := \emptyset, \qquad V_n := [0, 1] \setminus (C_n \cup V_{n-1}) \qquad \text{for} \quad n \geqslant 1,$$

then we have the following facts:

(1) The union of the V_n covers $[0, 1]$ up to a set of Lebesgue maesure 0.
(2) Each V_n consist of finitely many (exactly 2^{n-1} for $n \geqslant 1$) intervals.
(3) ψ is a polynomial of degree n on each interval of V_n.

6. Some further remarks

The basic objects of our study have been the predecessor sets

$$\mathcal{P}_T(a) = \left\{ n \in \mathbb{N} : T^k(n) = a \text{ for some } k \in \mathbb{N}_0 \right\},$$

of non-cyclic vertices $a \in \mathbb{N}$ of the Collatz graph Γ_T defined in II.1.7. We are especially interested in the asymptotics of the weighted counting function

$$W_{\mathcal{P}_T(a)} : \mathbb{N} \to \mathbb{Q}, \qquad W_{\mathcal{P}_T(a)}(x) = \frac{Z_a(x)}{x} = \frac{\left| \{ n \in \mathcal{P}_T(a) : n \leqslant x \} \right|}{x}.$$

With the notation of definition III.2.4, the point reached in chapter II is the estimate (corollary II.4.9)

$$(6.1) \qquad W_{\mathcal{P}_T(a)}(x) \geqslant \frac{\left| \mathcal{P}_T^n(a) \right|}{x} \geqslant \frac{s_n(a)}{x} \qquad \text{with} \quad n = \left\lfloor \log_2 \frac{x}{a} \right\rfloor.$$

A good result would be

$$(6.2) \qquad \liminf_{x \to \infty} W_{\mathcal{P}_T(a)}(x) > 0 \qquad \text{for} \quad a \not\equiv 0 \mod 3.$$

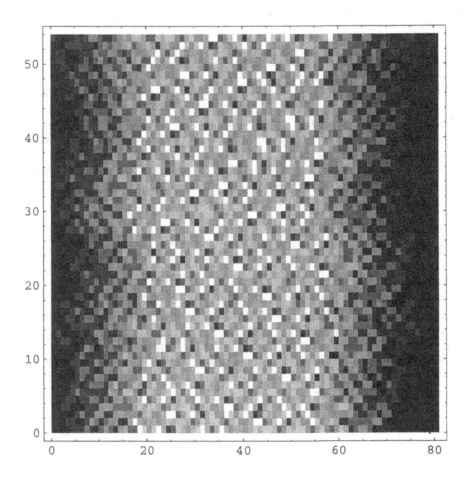

FIGURE 6. This is a density plot for the function $g_4(k, a)$. Here k runs from 0 to 80 in horizontal direction, and a is ordered according to the inverse Hensel code like in figure 4. Dark areas correspond to small values of g_4, and light areas correspond to large values of g_4. Together with figure 4, this density plot suggests that approximation (6.5) may be valid 'for large ℓ'.

If we could replace $s_n(a)$ by the 3-adic average $\bar{s}_n$ in the above estimate, this result would follow from theorem III.5.2.

The achievement of chapter IV may be considered as just one step into that direction. The definition of $s_n(a)$ combines with theorem 1.14 to give

$$(6.3) \qquad s_n(a) = \sum_{\ell=1}^{\infty} e_\ell(n + \lfloor \lambda \ell \rfloor, a) = \sum_{\ell=1}^{\infty} \sum_{j=0}^{n + \lfloor \lambda \ell \rfloor} p_\ell(n + \lfloor \lambda \ell \rfloor - j) \, g_\ell(j, a) \,.$$

In this formula, the only dependence on a is via the counting functions for small admissible vectors $g_\ell(j, a)$. By the renormalization lemma 2.14, we can write

$$(6.4) \qquad\qquad g_\ell(j, a) = \gamma_\ell^{-1} \, \varphi_\ell \left(\frac{j}{3^\ell}, a \right) \, .$$

The vague convergence of transition measures proved in theorem 4.1 nourishs the intuition that $\varphi_\ell \approx \varphi$ for large ℓ, where φ is the invariant density of the limiting transition probability given in theorem 5.1. In fact, if we could show that $\{\varphi_\ell : \ell \in \mathbb{N}\}$ is an equicontinuous family of functions on $\mathbb{B}_3 \times \mathbb{Z}_3^*$, we also could prove that the sequence (φ_ℓ) converges to φ in $L^1(\mathbb{B}_3 \times \mathbb{Z}_3^*, \varrho)$.

In order to show (6.2) using theorem III.5.2, it would suffice that the relative error $s_n(a)/\bar{s}_n$ is bounded away from zero for large n. In view of (6.3), this could be shown, if we can ensure that the approximation

$$(6.5) \qquad\qquad g_\ell(j, a) \approx \int_{\mathbb{Z}_3^*} g_\ell(j, b) \, d\nu_3(b)$$

is sufficiently close for large ℓ, and for arguments $j \approx n + \lfloor \lambda \ell \rfloor$.

On the other hand, we know by theorem III.2.7 that $s_n(a) = \infty$ on a dense set $S_\infty \subset \mathbb{Z}_3^*$. The conclusion is that we have to expect that the approximation in (6.5) is much better on $\mathbb{Z}_3^* \cap \mathbb{N}$ than on S_∞.

It would be interesting to see whether this expectation is related to an empirical observation of Applegate and Lagarias [**AL3**] (1995): They found that actual $3n + 1$ predecessor trees are significantly closer to the average than predicted by a stochastic branching process model.

CHAPTER V

MIXING AND PREDECESSOR DENSITY

The main subject in this chapter is the proof of the reduction theorem announced in the introduction. Logically, this proof only depends on the material of chapter II and the first two sections of chapter III. But heuristically, it is hard to imagine how to find the proof of theorem 2.1 without proving first something like theorem III.5.2. Moreover, the basic idea behind the notion 'locally covering triple' comes from the intuition gained in chapter IV. Indeed, a look at figure 5 nourishes the idea that multiple-step transition measures, or, dually, multiple-step dependence measures, would 'fill some columns' in figure 5. This idea finds its rigorous expression in the estimate of lemma 1.3,

$$
e_\ell(x, a) \geqslant 2 \cdot 3^{\ell-j-1} \int_{\mathbb{Z}_3^*} e_{\ell-j}(x - k, b)\, db = \binom{x - k + \ell - j}{\ell - j}.
$$

This estimate is valid if (j, k, ℓ) is a 'locally covering triple' at $a \in \mathbb{Z}_3^*$, which means, intuitively, that the j-step dependence measure at (x, a) 'fills the column' over the horizontal interval corresponding to $x - k$.

In the first section, the essential notion 'locally covering triple' is introduced, discussed, and related to the 3-adic geometry in the topological group $\mathbb{Z}_3^*$. In addition, the *normalized remainder map* is introduced. This map keeps book of the nonlinear parts of the $3n+1$ function accumulated when tracing back a finite portion of a $3n + 1$ trajectory. Certain images of this normalized remainder map can be described as sets of mixed power sums

$$
\mathcal{R}_{j,k} := \left\{ \sum_{i=0}^{j} 2^{\alpha_i} 3^i : j + k \geqslant \alpha_0 > \alpha_1 > \cdots > \alpha_j \geqslant 0 \right\} \subset \mathbb{Z}_3^*.
$$

The property 'locally covering triple' is also expressed as a distribution property in $\mathbb{Z}_3^*$ of such a set of mixed power sums.

The content of the second section is the proof of theorem 2.1 which connects estimates like that stated above to uniform lower bounds for $3n + 1$ predecessor counting functions. This proof uses well-known asymptotic properties of binomial coefficients.

In the final section, some consequences are discussed, and the proof of the reduction theorem is completed. The notion 'optimal sequence' at a non-cyclic number $a \in \mathbb{N} \setminus 3\mathbb{N}$ is defined and used to state a criterion for positive asymptotic density of a $3n + 1$ predecessor set $\mathcal{P}_T(a)$. The reduction theorem reduces the

UNIFORM SUB-POSITIVE PREDECESSOR DENSITY PROPERTY:

$$\liminf_{x \to \infty} \left(\inf_{a \not\equiv 0 \mod 3} \frac{Z_a(ax)}{x^\delta} \right) > 0 \qquad \text{for any } \delta \in \mathbb{R} \text{ satisfying } 0 < \delta < 1$$

to the

WEAK COVERING CONJECTURE FOR MIXED POWER SUMS. *Let $K(\ell) > 0$ be a function growing less than an exponential, i.e.,*

$$\lim_{\ell \to \infty} K(\ell)\, e^{-\gamma \ell} = 0 \qquad \text{for any constant } \gamma > 0 .$$

Then, for every $j, \ell \in \mathbb{N}$, the following implication holds:

$$|\mathcal{R}_{j-1,j}| \geqslant K(\ell) \cdot 3^\ell \quad \Longrightarrow \quad \mathcal{R}_{j-1,j} \text{ covers } \mathbb{Z}_3^* \text{ modulo } 3^\ell .$$

1. Locally covering triples

The basic counting functions $e_\ell : \mathbb{N}_0 \times \mathbb{N} \to \mathbb{N}_0$ have been given in definition II.4.1. According to corollary II.4.4, it is possible to contruct them inductively; given e_ℓ, there is a formal procedure producing $e_{\ell+1}$. Iterating this appropriately, lemma II.5.13 gives a formula computing e_ℓ out of $e_{\ell-j}$, for any fixed $j \in \{1, \dots, \ell - 1\}$.

Here we give a closer investigation of this topic, bearing in mind the intuition of the topological group $\mathbb{Z}_3^*$; these investigations are also related to j-step domains of dependence analoguous to those of definition IV.3.1. The aim is to find conditions for estimates of $e_\ell(k, a)$ in terms of binomial coefficients.

The basic estimate for locally covering triples. Recall the set $\mathcal{F}$ of feasible vectors (this means, non-negative integer vectors) and the notions of length and absolute value (cf. definition II.2.2). In II.2.9, we defined $v_s(a) \in \mathbb{Q}$ for each $s \in \mathcal{F}$ and $a \in \mathbb{Q}$. Here we shall meet sets of the following types (cf. definitions II.5.6 and II.5.11):

$$\mathcal{E}_{j,k}^*(a) = \left\{ s \in \mathcal{F} : \ell(s) = j,\ |s| = k,\ v_s(a) \in \mathbb{N},\ v_s(a) \not\equiv 0 \mod 3 \right\},$$

$$\mathcal{E}_{j,k}^{o*}(a) = \left\{ s \in \mathcal{F} : \ell(s) = j,\ |s| = k,\ v_s(a) \in \mathbb{N},\ v_s(a) \equiv 1, 5 \mod 6 \right\},$$

where $a \in \mathbb{Z}_3^*$ and $j, k \in \mathbb{N}_0$. We start with two definitions.

1.1. DEFINITION. Let $\ell \in \mathbb{N}_0$ be given. For two subsets $A, B \subset \mathbb{Z}_3$, we say that *$A$ covers B modulo 3^ℓ*, if $\pi_\ell(B) \subset \pi_\ell(A)$. Here π_ℓ denotes the *natural projection*

$$\pi_\ell : \mathbb{Z}_3 \to \mathbb{Z}_3/3^\ell \mathbb{Z}_3, \qquad \pi_\ell(x) = \left\{ x \mod 3^\ell \right\} .$$

1.2. DEFINITION. Let $j, k, \ell \in \mathbb{N}_0$ with $j < \ell$. We say that the triple (j, k, ℓ) is *locally covering at* $a \in \mathbb{Z}_3^*$, if

$$\{v_s(a) : s \in \mathcal{E}_{j,k}^*(a)\} \quad \text{covers} \quad \mathbb{Z}_3^* \quad \text{modulo} \quad 3^{\ell-j},$$

i.e., if the set $\{v_s(a) : s \in \mathcal{E}_{j,k}^*(a)\}$ contains a complete system of incongruent prime residues to modulus $3^{\ell-j}$.

1.3. LEMMA. *Let* $a \in \mathbb{N}$ *be not divisible by* 3, *let* $j, k, \ell \in \mathbb{N}_0$ *with* $j < \ell$, *and suppose that the triple* (j, k, ℓ) *is locally covering at* a. *Then*

$$e_\ell(x, a) \geqslant \binom{x - k + \ell - j}{\ell - j}.$$

PROOF. Like in definition III.3.1, let $A_{\ell-j}^*$ denote a complete set of incongruent prime residues to modulus $3^{\ell-j}$. Now the proof is a straightforward calculation using earlier results:

$$e_\ell(x, a) \geqslant \sum_{s \in \mathcal{E}_{j,k}^*(a)} e_{\ell-j}(x - i, v_s(a)) \qquad \text{by lemma II.5.14,}$$

$$\geqslant \sum_{b \in A_{\ell-j}^*} e_{\ell-j}(x - k, b) \qquad \text{by the hypothesis,}$$

$$\geqslant 2 \cdot 3^{\ell-j-1} \, \overline{e}_{\ell-j}(x - k) \qquad \text{by definition III.3.1,}$$

$$= \binom{x - k + \ell - j}{\ell - j} \qquad \text{by lemma III.3.2.} \quad \square$$

The normalized remainder map. The very question remaining after the last lemma is: When does the set $\{v_s(a) : s \in \mathcal{E}_{j,k}^*(a)\}$ cover a complete system of incongruent prime residues to modulus $3^{\ell-j}$?

A natural ansatz is to remember lemma II.2.13, which gives that $v_s(a) = c(s)a - r(s)$ where $c(s)$ is the coefficient and $r(s)$ the remainder of $s \in \mathcal{E}(a)$. According to definition II.2.12, the remainder of a feasible vector $s = (s_0, \ldots, s_\ell) \in \mathcal{F}$ is

$$r(s) = \sum_{j=0}^{\ell-1} \frac{2^{j+s_0+\cdots+s_j}}{3^{j+1}} = \frac{1}{3^\ell} \sum_{j=0}^{\ell-1} 2^{j+s_0+\cdots+s_j} \, 3^{\ell-j-1}.$$

In the last sum, all but one summands are divisible by 3. This implies $3^\ell r(s) \not\equiv 0$ mod 3, whence we can consider that number as an element of $\mathbb{Z}_3^*$.

1.4. DEFINITION. The *normalized remainder map* $R : \mathcal{F} \to \mathbb{Z}_3^*$ is given by

$$R(s) := 3^\ell r(s) = \sum_{j=0}^{\ell-1} \cdot 2^{j+s_0+\cdots+s_j} \cdot 3^{\ell-j-1} \quad \text{where } s = (s_0, \ldots, s_\ell).$$

The basic formula for normalized remainders follows from lemma II.2.13; let $a \in \mathbb{N}$ and $s \in \mathcal{E}_{j,k}(a)$, then

$$(1.1) \qquad v_s(a) = c(s)a - r(s) = \frac{1}{3^j}\left(2^{j+k}a - R(s)\right) .$$

We now determine to what extent this normalized remainder map is injective. From the defining formula, it is clear that $R(s)$ does not depend on the last component s_ℓ of the given vector $s = (s_0, \ldots, s_\ell)$. On the other hand, we shall see that two vectors $s, t \in \mathcal{F}$ with $\ell(s) = \ell(t)$, that give rise to the same normalized remainder $R(s) = R(t)$, can only differ in their last component. For the proof we use a lemma about mixed power sums.

1.5. LEMMA. *Let $j \in \mathbb{N}_0$, and let $\alpha_0 > \cdots > \alpha_j \geqslant 0$ and $\beta_0 > \cdots > \beta_j \geqslant 0$ denote two strictly decreasing finite sequences of integers. Then*

$$(\alpha_0, \ldots, \alpha_j) \neq (\beta_0, \ldots, \beta_j) \quad \Longrightarrow \quad \sum_{i=0}^{j} 2^{\alpha_i} 3^i \neq \sum_{i=0}^{j} 2^{\beta_i} 3^i .$$

PROOF. Suppose that $\alpha_i \neq \beta_i$ for some $i \in \{0, \ldots, \ell - 1\}$. Then there is a maximal index κ such that $\alpha_\kappa \neq \beta_\kappa$ and $\alpha_i = \beta_i$ when $\kappa + 1 \leqslant i \leqslant \ell - 1$, and the difference of the mixed power sums reads:

$$(1.2) \quad \sum_{i=0}^{j} 2^{\alpha_i} 3^i - \sum_{i=0}^{j} 2^{\beta_i} 3^i = \left(2^{\alpha_0} - 2^{\beta_0}\right) + 3\left(2^{\alpha_1} - 2^{\beta_1}\right) + \cdots + 3^\kappa \left(2^{\alpha_\kappa} - 2^{\beta_\kappa}\right) .$$

What is the highest power of 2 dividing this difference? The assumption that the α_i's and the β_i's are strictly decreasing implies

$$\min\{\alpha_0, \beta_0\} > \min\{\alpha_1, \beta_1\} > \cdots > \min\{\alpha_\kappa, \beta_\kappa\} \geqslant 0 .$$

This means that all but the last terms of the right hand side of (1.2) are divisible by $2^{\min\{\alpha_\kappa, \beta_\kappa\}+1}$. On the other hand, the highest power of 2 dividing the last term is $2^{\min\{\alpha_\kappa, \beta_\kappa\}}$. Hence

$$2^{\min\{\alpha_\kappa, \beta_\kappa\}+1} \nmid \sum_{i=0}^{j} 2^{\alpha_i} 3^i - \sum_{i=0}^{j} 2^{\beta_i} 3^i .$$

We conclude that these mixed power sums cannot be equal, which completes the proof. □

1.6. DEFINITION. Let $s = (s_0, \ldots, s_\ell) \in \mathcal{F}$ be a feasible vector. Then the *truncation* of s is defined by

$$s' := \begin{cases} (0) & \text{if } \ell(s) = 0 \\ (s_0, \ldots, s_{\ell-1}) & \text{otherwise.} \end{cases}$$

1.7. LEMMA. *Let $s, t \in \mathcal{F}$ satisfy $\ell(s) = \ell(t)$ and $R(s) = R(t)$. Then $s' = t'$.*

PROOF. If $\ell(s) = \ell(t) = 0$, there is nothing to prove. If $\ell(s) = \ell(t) = 1$, i. e. $s = (s_0, s_1)$ and $t = (t_0, t_1)$, we have $R(s) = 2^{s_0}$ and $R(t) = 2^{t_0}$. Hence, $R(s) = R(t)$ implies $s_0 = t_0$. From this, in turn, we infer $s' = (s_0) = (t_0) = t'$, which is the claim.

Now suppose that $\ell(s) = \ell(t) \geqslant 2$, and put $s = (s_0, \ldots, s_\ell)$ and $t = (t_0, \ldots, t_\ell)$. With the notation

$$\alpha_i := (\ell - 1 - i) + s_0 + \ldots + s_{\ell-1-i} \, ,$$
$$\beta_i := (\ell - 1 - i) + t_0 + \ldots + t_{\ell-1-i} \, , \qquad \text{for} \quad i = 0, \ldots, \ell - 1 \, ,$$

we have, clearly, $\alpha_0 > \alpha_1 > \cdots > \alpha_j \geqslant 0$ and $\beta_0 > \cdots > \beta_j \geqslant 0$. Definition 1.6 gives

$$R(s) = \sum_{i=0}^{\ell-1} 2^{\alpha_i} 3^i \qquad \text{and} \qquad R(t) = \sum_{i=0}^{\ell-1} 2^{\beta_i} 3^i \, .$$

It is plain that $s' \neq t'$ implies $\alpha_i \neq \beta_i$ for some $i \leqslant \ell - 1$. Hence, an application of lemma 1.5 completes the proof. $\square$

1.8. COROLLARY & NOTATION. *For $j, k \in \mathbb{N}_0$ denote*

$$\mathcal{F}_{j,k} := \{ s = (s_0, \ldots, s_j) \in \mathcal{F} : \ell(s) = j, |s| = k \} \qquad \text{and}$$
$$\mathcal{R}_{j,k} := \left\{ \sum_{i=0}^j 2^{\alpha_i} 3^i : j + k \geqslant \alpha_0 > \alpha_1 > \cdots > \alpha_j \geqslant 0 \right\}.$$

Then $\quad \mathcal{R}_{j,k} = R(\mathcal{F}_{j+1,k}) \quad$ *and* $\quad |\mathcal{R}_{j,k}| = \dbinom{j + k + 1}{j + 1}$.

PROOF. The map

$$\{ (\alpha_0, \ldots, \alpha_j) : j + k \geqslant \alpha_0 > \alpha_1 > \cdots > \alpha_j \geqslant 0 \} \quad \longrightarrow \quad \mathcal{F}_{j+1,k}$$

defined by $s_0 := \alpha_j$, $s_i := \alpha_{j-i} - \alpha_{j-i+1} - 1$ for $i = 1, \ldots, j$, and $s_{j+1} := j + k - \alpha_0$, is easily checked to be a bijection. This proves the first equation.

To check the computation of the cardinality, observe that a $(j + 1)$-tuple $(\alpha_0, \ldots, \alpha_j)$ with the property $j + k > \alpha_0 > \alpha_1 > \cdots > \alpha_j \geqslant 0$ is unambiguously given by the choice of $j + 1$ elements out of the set $\{0, \ldots, j + k\}$ of $j + k + 1$ elements. $\square$

3-adic balls and spheres. The next step is to rewrite definition II.2.15 in terms of this normalized remainder map. Let $s \in \mathcal{F}$ be a feasible vector. Then, by lemma II.2.13,

$$s \in \mathcal{E}_{j,k}(a) \quad \Longleftrightarrow \quad v_s(a) \in \mathbb{N} \quad \Longleftrightarrow \quad 2^{j+k} a \equiv R(s) \mod 3^j \, .$$

Moreover, by definition II.5.6, we have the equivalences

$$s \in \mathcal{E}_{j,k}^*(a) \quad \Longleftrightarrow \quad v_s(a) \in \mathbb{N} \quad \text{and} \quad v_s \not\equiv 0 \mod 3$$

$$(1.3) \qquad \Longleftrightarrow \quad 2^{j+k} a \equiv R(s) \mod 3^j \quad \text{but} \quad 2^{j+k} a \not\equiv R(s) \mod 3^{j+1} \, .$$

This motivates a connection to balls and spheres in the 3-adic metric d_3, as $d_3(b, c) = 3^{-j}$ where 3^j is the highest power of 3 dividing $b - c$ (cf. definition III.1.2).

1.9. DEFINITION. Let $b \in \mathbb{Z}_3^*$ and $j \in \mathbb{N}$. Taking b as center and 3^{-j} as radius, we define

the *3-adic ball*: $B(b, 3^{-j}) := \left\{ c \in \mathbb{Z}_3^* : d_3(b, c) \leqslant 3^{-j} \right\}$,

and the *3-adic sphere*: $S(b, 3^{-j}) := \left\{ c \in \mathbb{Z}_3^* : d_3(b, c) = 3^{-j} \right\}$.

1.10. REMARK. It is worth to write these sets in the residue class notation introduced in definition III.1.7:

$$B(b, 3^{-j}) = \left\{ b \mod 3^j \right\} .$$

The 3-adic sphere can be written in different ways:

$$
\begin{aligned}
S(b, 3^{-j}) &= B(b, 3^{-j}) \setminus B(b, 3^{-j-1}) \\
&= \left\{ b \mod 3^j \right\} \setminus \left\{ b \mod 3^{j+1} \right\} \\
&= \left\{ c \in \mathbb{Z}_3^* : c \equiv b \mod 3^j \text{ and } c \not\equiv b \mod 3^{j+1} \right\} \\
&= \left\{ b + 3^j \mod 3^{j+1} \right\} \cup \left\{ b - 3^j \mod 3^{j+1} \right\} \\
&= B(b + 3^j, 3^{-j-1}) \cup B(b - 3^j, 3^{-j-1}) .
\end{aligned}
$$

Note that a ball $B(b, 3^{-j})$ is both open and closed in the 3-adic topology. Especially, the 3-adic sphere $S(b, 3^{-j})$ is *not* the topological boundary of the 3-adic ball $B(b, 3^{-j})$; indeed, that topological boundary is empty. The situation is even more contra-intuitive: the 3-adic volume of the sphere $S(b, 3^{-j})$ is just twice the 3-adic volume of the ball $B(b, 3^{-j})$.

We now start to reformulate the notion "locally covering triple" in terms of the 3-adic metric on the topological group $\mathbb{Z}_3^*$ of invertible 3-adic integers. In the next lemma, we take advantage of the fact that $\mathbb{Z}_3^*$ is a topological group in the following way: Any $b \in \mathbb{Z}_3^*$ is invertible in $\mathbb{Z}_3^*$; we write $b^{-1} \in \mathbb{Z}_3^*$ for its multiplicative inverse. Now for any $a, b \in \mathbb{Z}_3^*$, we have the set equation

(1.4) $b^{-1} S\left(b\, a, 3^{-j} \right) = S\left(a, 3^{-j} \right)$.

1.11. LEMMA. *Let $a \in \mathbb{Z}_3^*$ and $j, k, \ell \in \mathbb{N}_0$ with $j < \ell$. Then the following assertions are equivalent:*

(a) *The triple (j, k, ℓ) is locally covering at a.*
(b) *For any $b \in \mathbb{Z}_3^*$ with $b \equiv 2^{j+k} a \mod 3^j$ and $b \not\equiv 2^{j+k} a \mod 3^{j+1}$, there is an $s \in \mathcal{E}_{j,k}^*(a)$ such that $R(s) \equiv b \mod 3^\ell$.*
(c) *The set $2^{-j-k} R\left(\mathcal{E}_{j,k}^*(a) \right)$ covers $S(a, 3^{-j})$ modulo 3^ℓ.*
(d) *The set $2^{-j-k} R\left(\mathcal{F}_{j,k} \right)$ covers $S(a, 3^{-j})$ modulo 3^ℓ.*

PROOF. (a) $\Rightarrow$ (b): From (a) we conclude that $v_s(a)$ hits any prime residue class to modulus $3^{\ell-j}$, when s runs through the set $\mathcal{E}_{j,k}^*(a)$. By formula (1.1), we infer that the difference $2^{j+k}a - R(s)$ must hit any residue class $x \mod 3^\ell$ which satisfies both

$$x \equiv 0 \mod 3^j \qquad \text{and} \qquad x \not\equiv 0 \mod 3^{j+1}.$$

(b) $\Rightarrow$ (c): Using definition 1.1, we obtain that (b) means that the set $R\left(\mathcal{E}_{j,k}^*(a)\right)$ covers the 3-adic sphere $S\left(2^{j+k}a, 3^{-j}\right)$ modulo 3^ℓ. Now (c) follows from (1.4).

(c) $\Leftrightarrow$ (d): The equivalence (1.3) implies:

$$\mathcal{E}_{j,k}^*(a) = \left\{ s \in \mathcal{F}_{j,k} : R(s) \in S\left(2^{j+k}a, 3^{-j}\right) \right\}.$$

Hence, (c) $\Leftrightarrow$ (d) follows from (1.4).

(c) $\Rightarrow$ (a): It is plain that, for any fixed $x \in \mathbb{Z}_3^*$, the bijection

$$q_j : S\left(x, 3^{-j}\right) \to \mathbb{Z}_3^*, \qquad q_j(y) := \frac{x - y}{3^j}$$

induces a bijection $\widetilde{q}_j : \pi_\ell\left(S(x, 3^{-j})\right) \to \pi_{\ell-j}(\mathbb{Z}_3^*)$ which satisfies $\widetilde{q}_j \circ \pi_\ell = \pi_{\ell-j} \circ q_j$.

Now fix $x := 2^{j+k}a$. Then we have $v_s(a) = q_j(R(s))$ for any vector $s \in \mathcal{E}_{j,k}^*(a)$, and we obtain using (c):

$$\begin{aligned}
\pi_{\ell-j}\left\{ v_s(a) : s \in \mathcal{E}_{j,k}^*(a) \right\} &= \pi_{\ell-j} \circ q_j \left\{ R(s) : s \in \mathcal{E}_{j,k}^*(a) \right\} \\
&= \widetilde{q}_j \circ \pi_\ell \left(R(\mathcal{E}_{j,k}^*(a)) \right) \\
&= \widetilde{q}_j \left(\pi_\ell \left(S(2^{j+k}a, 3^{-j}) \right) \right).
\end{aligned}$$

Now claim (a) follows from the fact the $\widetilde{q}_j$ is a bijection. $\square$

Globally covering triples. It will turn out to be interesting to see what happens if we "globalize" definition 1.2:

1.12. DEFINITION. Let $j, k, \ell \in \mathbb{N}_0$ with $j < \ell$. We say that the triple (j, k, ℓ) is *globally covering*, if it is locally covering at each $a \in \mathbb{Z}_3^*$.

1.13. LEMMA. *Let $j, k, \ell \in \mathbb{N}_0$ with $j < \ell$. Then the following assertions are equivalent:*

(a) *The triple (j, k, ℓ) is globally covering.*
(b) *The set $R\left(\mathcal{F}_{j,k}\right)$ covers $\mathbb{Z}_3^*$ modulo 3^ℓ.*

PROOF. The equivalence follows from part (a) $\Leftrightarrow$ (d) of lemma 1.11, as the group $\mathbb{Z}_3^*$ can be written as a (non-disjoint) union of spheres of radius 3^{-j}. $\square$

2. A predecessor density criterion

Theorem III.2.5 proved a link between the asymptotic behaviour of the *estimating series* $s_n(a)$, as $n \to \infty$, and a density estimates for the $3n+1$ *predecessor set* $\mathcal{P}_T(a)$, for each $a \in \mathbb{N}$ which is non-cyclic in the Collatz graph Γ_T. Recall from definition III.2.4 that the estimating series is given by

$$s_n(a) = \sum_{\ell=1}^{\infty} e_\ell(n + \lfloor \lambda\ell \rfloor, a),$$

with the constant $\lambda = \log_2(\frac{3}{2})$. Then theorem III.2.5 says that, for a real number $\beta \geq 1$,

$$\liminf_{n\to\infty} \left(\inf_{a\in A} \frac{s_n(a)}{\beta^n} \right) > 0 \quad \Longrightarrow \quad \liminf_{n\to\infty} \left(\inf_{a\in A} \frac{Z_a(ax)}{x^{\log_2 \beta}} \right) > 0.$$

Now we are ready to prove a link between estimates for the counting function as given in lemma 1.3 and estimates for the asymptotic behaviour of $s_n(a)$. The strategy for the proof is to provide a appropriately refined elaboration of the idea of the proof of theorem 5.2.

2.1. THEOREM. *Let $A \subset \mathbb{Z}_3^*$ be a given subset, and let $(j_\ell)_{\ell\in\mathbb{N}}$ and $(k_\ell)_{\ell\in\mathbb{N}}$ denote two sequences in $\mathbb{N}_0$ with $j_\ell < \ell$ and*

$$e_\ell(k,a) \geq \binom{k - k_\ell + \ell - j_\ell}{\ell - j_\ell} \qquad \text{for all} \quad a \in A \quad \text{and} \quad k, \ell \in \mathbb{N};$$

here the binomial coefficient equal zero if $k < k_\ell$. In addition, suppose that the limits

$$\eta := \lim_{\ell\to\infty} \frac{j_\ell}{\ell} \qquad \text{and} \qquad \xi := \lim_{\ell\to\infty} \frac{k_\ell}{\ell}$$

exist and satisfy $\eta < 1$ and $2 - \alpha + \xi - \eta > 0$ with $\alpha = \log_2 3$. Then

$$\liminf_{n\to\infty} \left(\inf_{a\in A} \frac{s_n(a)}{\beta^n} \right) > 0 \qquad \text{whenever} \quad \log_2 \beta < \frac{2(1 - \eta)}{2 - \alpha + \xi - \eta}.$$

Moreover, if the above quotients converge so fast that $|\eta\ell - j_\ell|$ and $|\xi\ell - k_\ell|$ remain bounded for all $\ell \in \mathbb{N}$, then

$$\liminf_{n\to\infty} \left(\inf_{a\in A} \frac{s_n(a)}{\beta^n} \right) > 0 \qquad \text{with} \quad \log_2 \beta = \frac{2(1 - \eta)}{2 - \alpha + \xi - \eta}.$$

PROOF. The main idea of the proof is first to use the estimate in the assumption and then proceed along the lines of the proof of theorem III.5.2. With the constants $\alpha = \lambda + 1 = \log_2 3$, this gives

$$(2.1) \quad \inf_{a\in A} \left(\frac{s_n(a)}{\beta^n} \right) = \inf_{a\in A} \left(\sum_{\ell=1}^{\infty} \frac{e_\ell(n + \lambda\ell, a)}{\beta^n} \right) \geq \sum_{\ell=1}^{\infty} \frac{1}{\beta^n} \binom{n + \lfloor \alpha\ell \rfloor - j_\ell - k_\ell}{\ell - j_\ell}$$

We shall use the notation

(2.2) $$\tilde{n}_\ell := n + \lfloor \alpha\ell \rfloor - j_\ell - k_\ell \, .$$

Lemma III.5.1 asserts that the quantities $c(\ell)$ defined by

(2.3) $$\frac{1}{2^{\tilde{n}_\ell}} \binom{\tilde{n}_\ell}{\ell - j_\ell} = c(\ell) \cdot \left(\frac{2}{\pi \tilde{n}_\ell} \right)^{1/2} \exp\left(-\frac{\tilde{x}_\ell(n)^2}{2} \right) \, ,$$

with the notation $\tilde{x}_\ell(n) := (2(\ell - j_\ell) - \tilde{n}_\ell)/\sqrt{\tilde{n}_\ell}$, satisfy

(2.4) $$\lim_{\ell \to \infty} c(\ell) = 1 \, , \qquad \text{provided } |\tilde{x}_\ell(n)| \leqslant 1 \text{ for large } \ell \, .$$

We shall show that the sequence of sets

(2.5) $$\widetilde{\Delta}_n := \{\ell \in \mathbb{N} : \tilde{x}_\ell(n)^2 \leqslant 1\} = \{\ell \in \mathbb{N} : (2(\ell - j_\ell) - \tilde{n}_\ell)^2 \leqslant \tilde{n}_\ell\}$$

satisfies the two conditions:

(2.6) $$\lim_{n \to \infty} \left(\inf \widetilde{\Delta}_n \right) = \infty \, ,$$

and there is a constant $c_1 > 0$ such that

(2.7) $$|\widetilde{\Delta}_n| \geqslant c_1 \cdot \tilde{n}_\ell^{1/2} \qquad \text{for each } \ell \in \widetilde{\Delta}_n \, .$$

This will lead through (2.3) and (2.4) to an estimate for (2.1), for appropriate β.

With the notations

$$\delta_\ell := 1 - \frac{j_\ell}{\ell} \qquad \text{and} \qquad \gamma_\ell := \frac{\lfloor \alpha\ell \rfloor - j_\ell - k_\ell}{\ell} \, ,$$

we have $\ell - j_\ell = \delta_\ell \ell$, and we derive $\tilde{n}_\ell = n + \gamma_\ell \ell$ from (2.2). Now

$$\left(\tilde{n}_\ell - 2(\ell - j_\ell) \right)^2 - \tilde{n}_\ell = (n + \gamma_\ell \ell - 2\delta_\ell \ell)^2 - n - \gamma_\ell \ell$$
$$= (n - \sigma_\ell \ell)^2 - n - \gamma_\ell \ell$$

with the abbreviation $\sigma_\ell := 2\delta_\ell - \gamma_\ell$,

(2.8) $$= \sigma_\ell^2 \ell^2 - (2\sigma_\ell n + \gamma_\ell) \ell + n^2 - n$$

At this point we insert the assumption that the sequences $(j_\ell/\ell)_{\ell \in \mathbb{N}}$ and $(k_\ell/\ell)_{\ell \in \mathbb{N}}$ converge. This implies that σ_ℓ and γ_ℓ vary, for large ℓ, in a small neighbourhood of their limits σ_∞ and γ_∞, respectively. Hence it makes sence to consider (2.8) as a perturbed quadratic form in ℓ. The "discriminant" of this quadratic form in ℓ would be

$$(2.9) \qquad \widetilde{D}_{\ell,n} := (2\sigma_\ell n + \gamma_\ell)^2 - 4\sigma_\ell^2(n^2 - n) = 8\sigma_\ell \delta_\ell\, n + \gamma_\ell^2 \,,$$

as $\sigma_\ell + \gamma_\ell = 2\delta_\ell$. We use this "discriminant" to rewrite $\widetilde{\Delta}_n$ from (2.5):

$$\widetilde{\Delta}_n = \{\ell \in \mathbb{N} : \sigma_\ell^2\,\ell^2 - (2\sigma_\ell n + \gamma_\ell)\,\ell + n^2 - n \leqslant 0\}$$

$$(2.10) \qquad = \left\{\ell \in \mathbb{N} : \frac{2\sigma_\ell + \gamma_\ell - \widetilde{D}_{\ell,n}^{1/2}}{2\sigma_\ell^2} \leqslant \ell \leqslant \frac{2\sigma_\ell + \gamma_\ell + \widetilde{D}_{\ell,n}^{1/2}}{2\sigma_\ell^2}\right\}\,.$$

If ℓ is sufficiently large, then $\sigma_\ell \approx \sigma_\infty$, $\gamma_\ell \approx \gamma_\infty$, and $\widetilde{D}_{\ell,n} \approx \widetilde{D}_{\infty,n}$, and $\widetilde{\Delta}_n$ is approximately an interval of length $\widetilde{D}_{\infty,n}^{1/2}/\sigma_\infty^2$ centered at

$$(2.11) \qquad \ell(n) := \frac{2\sigma_\infty n + \gamma_\infty}{2\sigma_\infty^2} = \frac{n}{\sigma_\infty} + \frac{\gamma_\infty}{2\sigma_\infty^2}\,.$$

Note that this implies (2.6), provided $\sigma_\infty > 0$.

Let us state the essential points a bit more explicitly. First, the limiting quantities with index ∞ are easily computed using the limits η and ξ:

$$\delta_\infty = \lim_{\ell \to \infty}\left(1 - \frac{j_\ell}{\ell}\right) = 1 - \eta\,,$$

$$\gamma_\infty = \lim_{\ell \to \infty}\frac{\lfloor \alpha\ell \rfloor - j_\ell - k_\ell}{\ell} = \alpha - \eta - \xi\,,$$

$$\sigma_\infty = 2\delta_\infty - \gamma_\infty = 2 - \alpha + \xi - \eta\,.$$

Note that the conditions $\eta < 1$ and $2 - \alpha - \eta + \xi > 0$ just mean $\delta_\infty > 0$ and $\sigma_\infty > 0$, respectively. Hence we have proved (2.6).

Let us denote

$$\varepsilon_\ell := \max\left\{\left|\eta - \frac{j_\ell}{\ell}\right|, \left|\xi - \frac{k_\ell}{\ell}\right|\right\}\,.$$

With the usual O-notation asserting that the quotient $O(\varepsilon_\ell)/\varepsilon_\ell$ remains bounded for $\ell \to \infty$, we infer from (2.9) and our limit investigations the formula

$$(2.12) \qquad \widetilde{D}_{\ell,n} = \big(8\sigma_\infty\delta_\infty + O(\varepsilon_\ell)\big)n\,.$$

Inserting this into (2.10), we obtain with (2.11)

$$(2.13) \qquad \ell = (1 + O(\varepsilon_\ell))\,\ell(n) = n\left(\frac{1}{\sigma_\infty} + O(\varepsilon_\ell)\right) \qquad \text{for} \quad \ell \in \widetilde{\Delta}_n\,.$$

Moreover, for $\ell \in \widetilde{\Delta}_n$ we have the equations

$$(2.14) \quad \tilde{n}_\ell = n + \gamma_\ell\ell = n + (\gamma_\infty + O(\varepsilon_\ell))\,n\left(\frac{1}{\sigma_\infty} + O(\varepsilon_\ell)\right) = n\left(\frac{2\delta_\infty}{\sigma_\infty} + O(\varepsilon_\ell)\right)\,.$$

Together with (2.12), this implies (2.7).

Next we seek an appropriate β. As suggested by (2.3), we should have that β^n is not essentially larger than $2^{\tilde{n}_\ell}$ for $\ell \in \tilde{\Delta}_n$. Formally, we should know that there is a constant $c_4 > 0$ such that

(2.15) $\beta^n \leqslant c_4 \, 2^{\tilde{n}_\ell}$ for sufficiently large $n \in \mathbb{N}$ and $\ell \in \tilde{\Delta}_n$.

Taking into account (2.14), we infer that this is valid if

(2.16) $\log_2 \beta - \dfrac{2\delta_\infty}{\sigma_\infty} \leqslant \dfrac{\log_2 c_4}{n} + O(\varepsilon_\ell)$ for large $n \in \mathbb{N}$ and $\ell \in \tilde{\Delta}_n$.

Because $\varepsilon_\ell \to 0$ for $\ell \to \infty$, this relation holds for sufficiently large n whenever

$$\log_2 \beta < \frac{2\delta_\infty}{\sigma_\infty} = \frac{2(1 - \eta)}{2 - \alpha + \xi - \eta} \, .$$

Moreover, if $|\eta\ell - j_\ell|$ and $|\xi\ell - k_\ell|$ remain bounded for all $\ell \in \mathbb{N}$, we have $\varepsilon_\ell = O(\ell^{-1})$. From (2.13) we derive

$$\varepsilon_\ell = O\left(\frac{1}{\ell}\right) = O\left(\frac{1}{n}\right)   \text{for}   \ell \in \tilde{\Delta}_n \, .$$

Hence (2.16) gives that (2.15) is fulfilled if

$$\log_2 \beta - \frac{2\delta_\infty}{\sigma_\infty} \leqslant \frac{\log_2 c_4}{n} + O\left(\frac{1}{n}\right)   \text{for all}   n \in \mathbb{N} \, .$$

This, in turn, can be guaranteed by choosing a sufficiently large constant c_4, even if

$$\log_2 \beta = \frac{2\delta_\infty}{\sigma_\infty} = \frac{2(1 - \eta)}{2 - \alpha + \xi - \eta} \, .$$

Now we are ready to do the final calculation, which are quite similar to those at the end of the proof of theorem III.5.2:

$$\inf_{a \in A} \left(\frac{s_n(a)}{\beta^n}\right)$$

$$\geqslant \sum_{\ell=1}^{\infty} \frac{1}{\beta^n} \binom{\tilde{n}_\ell}{\ell - j_\ell}   \text{by (2.1)},$$

$$\geqslant c_4^{-1} \sum_{\ell \in \tilde{\Delta}_n} \frac{1}{2^{\tilde{n}_\ell}} \binom{\tilde{n}_\ell}{\ell - j_\ell}   \text{by (2.15) for large } n,$$

$$= c_4^{-1} \sum_{\ell \in \tilde{\Delta}_n} c(\ell) \cdot \left(\frac{2}{\pi \tilde{n}_\ell}\right)^{1/2} \exp\left(-\frac{\tilde{x}_\ell(n)^2}{2}\right)   \text{by (2.3)},$$

$$\geqslant c_4^{-1} \sqrt{\frac{2}{\pi e}} \sum_{\ell \in \tilde{\Delta}_n} c(\ell) \, \tilde{n}_\ell^{-1/2}   \text{by (2.5)},$$

$$\geqslant c_5 \cdot \min_{\ell \in \tilde{\Delta}_n} c(\ell) \cdot |\tilde{\Delta}_n| \left(\max_{\ell \in \tilde{\Delta}_n} \tilde{n}_\ell\right)^{-1/2}   \text{with } c_5 = c_4^{-1} \sqrt{\frac{2}{\pi e}},$$

$$\geqslant c_5 \cdot c_1 \cdot \min_{\ell \in \tilde{\Delta}_n} c(\ell)   \text{by (2.7)}.$$

By (2.4), the latter converges to $c_5 c_1 > 0$ as n tends to ∞, which completes the proof. $\square$

3. Consequences

This section gives some consequences of theorem 2.1. We start by defining the notion "optimal sequence" at some fixed $a \in \mathbb{Z}_3^*$. This notion is defined in such a way that existence of an optimal sequence at $a \in \mathbb{N} \cap \mathbb{Z}_3^*$ just gives the conditions which allow theorem 2.1 to imply that the $3n+1$ predecessor set $\mathcal{P}_T(a)$ has positive asymptotic density. Thus, we are led to the deep problem whether it is possible that the family of $3n+1$ predecessor sets $\{\mathcal{P}_T(a) : a \in \mathbb{N}, a \not\equiv 0 \bmod 3\}$ has "uniform positive density". It is shown that it is impossible that this family has both a linear uniform lower bound and a linear uniform upper bound; this shows that the uniform positive density problem is intimately connected to the dynamics of $3n + 1$ iterations.

Next we prove by a counting argument that it is impossible to satisfy the sufficient condition for positive asymptotic density "uniformly" for all numbers $a \not\equiv 0 \bmod 3$. Finally, we formulate a covering conjecture for mixed powers sums, and we prove the reduction theorem using a variation of that counting argument.

A sufficient condition for positive density. Let us figure out in detail what are the conditions which would enable us to prove, using theorem 2.1, that a $3n + 1$ predecessor set has positive lower asymptotic density. Recall the very definition of the $3n+1$ funktion T in section I.1, and the definition of predecessor sets:

$$\mathcal{P}_T(a) = \{b \in \mathbb{N} : \text{there is a } k \in \mathbb{N}_0 \text{ with } T^k(b) = a\} .$$

The counting function (cf. definition II.1.15) is

$$Z_a(x) = |\{\nu \in \mathcal{P}_T(a) : \nu \leqslant x\}| ,$$

and the set $\mathcal{P}_T(a)$ is said to have *positive lower asymptotic density*, if the weighted counting function has positive *limes inferior*:

$$\liminf_{x \to \infty} W_{\mathcal{P}_T(a)}(x) = \liminf_{x \to \infty} \frac{Z_a(x)}{x} > 0 .$$

To state the sufficient condition, we introduce the following notion.

3.1. DEFINITION. Let $a \in \mathbb{Z}_3^*$. A sequence of pairs $(j_\ell, k_\ell)_{\ell \in \mathbb{N}}$ is called an *optimal sequence at a*, if the following conditions are satisfied:

  (i) For each $\ell \in \mathbb{N}$, the triple (j_ℓ, k_ℓ, ℓ) is locally covering at a.
  (ii) The limits $\eta := \lim_{\ell \to \infty} (j_\ell/\ell)$ and $\xi := \lim_{\ell \to \infty} (k_\ell/\ell)$ exist.
  (iii) These limits fulfill $\eta < 1$ and $\xi + \eta = \alpha$ with $\alpha = \log_2 3$.
  (iv) The set $\{|\eta\ell - j_\ell|, |\xi\ell - k_\ell| : \ell \in \mathbb{N}\}$ is bounded.

3.2. LEMMA. *Let $a \in \mathbb{N}$ be non-cyclic and satisfying $a \not\equiv 0 \bmod 3$, and suppose that there is an optimal sequence at a. Then there is a constant $c > 0$ such that the $3n + 1$ predecessor set satisfies*

$$Z_a(x) \geqslant c\, x \qquad \text{for sufficiently large } x.$$

PROOF. Let $(j_\ell, k_\ell)_{\ell \in \mathbb{N}}$ be an optimal sequence at $a \in \mathbb{N} \setminus 3\mathbb{N} \subset \mathbb{Z}_3^*$, and denote by η and ξ the limits as in definition 3.1. We want to apply theorem 2.1 and theorem III.2.5. By the latter, and in order to get positive asymptotic density, we have to ensure

$$\liminf_{n \to \infty} \frac{s_n(a)}{\beta^n} > 0 \qquad \text{with} \quad \log_2 \beta = 1 \quad (\text{which means } \beta = 2).$$

So we have to check what are the conditions on the limits ξ and η which admit the choice $\beta = 2$ in theorem 2.1:

$$\log_2 \beta = 1 \quad \Longleftrightarrow \quad 2(1 - \eta) = 2 - \alpha + \xi - \eta \quad \Longleftrightarrow \quad \xi + \eta = \alpha.$$

But this condition is satisfied when (j_ℓ, k_ℓ) is an optimal sequence at a. By this, the condition $\eta < 1$, and the condition on the speed of convergence (see (iv) of definition 3.1), we infer from theorem 2.1 that

$$\liminf_{n \to \infty} \frac{s_n(a)}{2^n} > 0.$$

Now theorem III.2.5 completes the proof. $\square$

Uniform positive density. Let's suppose for the moment that we could satisfy the conditions of definition 3.1 uniformly for all $a \in \mathbb{Z}_3^*$. An inspection of the proofs of theorem 2.1 and theorem III.2.5 shows that this would imply that the family $\{\mathcal{P}_T(a) : a \in \mathbb{N} \setminus 3\mathbb{N}\}$ admits a uniform lower bound(in the sense of definition II.6.1) of the form $\varphi(\xi) = c\xi$ for some real constant $c > 0$. In other words, the $3n + 1$ predecessor sets would have the following

3.3. UNIFORM POSITIVE DENSITY PROPERTY. *There are real constants $c > 0$ and $\xi_0 > 0$ such that*

$$Z_a(x) \geqslant \frac{c\,x}{a} \qquad \text{for} \quad a \in \mathbb{N}, \quad a \not\equiv 0 \mod 3, \quad \frac{x}{a} \geqslant \xi_0,$$

Should we expect that the $3n + 1$ predecessor sets have this uniform positive density property? Before answering immediately 'yes, we should expect that,' consider the following lemma. It states that it cannot be true that the family $\{\mathcal{P}_T(a) : a \in \mathbb{N} \setminus 3\mathbb{N} \text{ non-cyclic}\}$ admits both a linear uniform lower bound and a linear uniform upper bound.

3.4. LEMMA. *The following assertions cannot be true simultaneously.*

(a) *There is a real number $c_1 > 0$ such that $\varphi_1(\xi) = c_1 \xi$ is a uniform lower bound for the family $\{\mathcal{P}_T(a) : a \in \mathbb{N} \setminus 3\mathbb{N} \text{ non-cyclic}\}$.*

(b) *There is a real number $c_2 > 0$ such that $\varphi_2(\xi) = c_2 \xi$ is a uniform upper bound for the family $\{\mathcal{P}_T(a) : a \in \mathbb{N} \setminus 3\mathbb{N} \text{ non-cyclic}\}$.*

PROOF. Suppose that both assertions (a) and (b) are true, and suppose that we have numbers $a, b \in U$ such that b is a predecessor of a, which means $\mathcal{P}_T(b) \subset \mathcal{P}_T(a)$. Then there are constants $c_2 > c_1 > 0$ and $\xi_0 > 0$ such that

$$\frac{c_2 x}{a} \geqslant Z_a(x) \geqslant Z_b(x) \geqslant \frac{c_1 x}{b} \qquad \text{for} \qquad \min\left\{\frac{x}{a}, \frac{x}{b}\right\} \geqslant \xi_0 .$$

Here the first inequality follows from (b), and the last inequality is obtained from (a). This chain of inequalities implies

$$(3.1) \qquad \frac{a}{b} \leqslant \frac{c_2}{c_1} \qquad \text{for all non-cyclic} \quad a, b \in \mathbb{N} \setminus 3\mathbb{N} \quad \text{with} \quad b \in \mathcal{P}_T(a) .$$

We construct appropriate a, b such that their quotient a/b is arbitrarily large. More precisely, to each *odd* number $k > 0$ there are numbers a_k, b_k such that

(i) $T^k(b_k) = a_k$, where T is the $3n + 1$ function I.(1.2),

(ii) $a_k/b_k > (3/2)^k$,

(iii) $a_k, b_k \not\equiv 0 \mod 3$, and

(iv) a_k and b_k are non-cyclic.

To this end, observe that $T^k\left(2^k z - 1\right) = 3^k z - 1$ for every $z \in \mathbb{N}$. As k is odd, the choice $a_k := 3^k z_k - 1$, $b_k := 2^k z_k - 1$ fulfills (i)–(iii) for arbitrary numbers $z_k \in \mathbb{N}$. To ensure (iv), it suffices to choose a_k non-cyclic, as predecessors of non-cyclic numbers are also non-cyclic. To achieve this, observe

$$a_k = 3^k z_k - 1 = 2^m \qquad \Longleftrightarrow \qquad z_k = \frac{2^m + 1}{3^k} .$$

There is such an integer z_k whenever $m \equiv 3^{k-1} \mod 3^k$, and $a_k = 2^m$ is clearly non-cyclic, if $m > 2$. This implies that (3.1) cannot be true. $\square$

Non-existence of globally optimal sequences. Having defined (in definition 3.1) optimal sequences at a fixed point $a \in \mathbb{Z}_3^*$, it is natural to call a sequence of pairs (j_ℓ, k_ℓ) *globally optimal*, if it is an optimal sequence at each point of $\mathbb{Z}_3^*$. But such objects do not exist.

3.5. THEOREM. *There is no globally optimal sequence.*

PROOF. Suppose that $(j_\ell, k_\ell)_{\ell \in \mathbb{N}}$ is a globally optimal sequence, and use the notation

$$\eta = \lim_{\ell \to \infty} \frac{j_\ell}{\ell} \qquad \text{and} \qquad \xi = \lim_{\ell \to \infty} \frac{k_\ell}{\ell} ,$$

where the limits exist by (ii) of definition 3.1. Note that condition (iii) of this definition gives $\xi + \eta = \alpha = \log_2 3$.

By (i) of definition 3.1, we know that each triple (j_ℓ, k_ℓ, ℓ) is locally covering at each point $a \in \mathbb{Z}_3^*$. Then, by definition 1.12, the triple (j_ℓ, k_ℓ, ℓ) is globally covering for each $\ell \in \mathbb{N}$. We infer by lemma 1.13 that, for any $\ell \in \mathbb{N}$, the set $R(\mathcal{F}_{j_\ell, k_\ell})$ covers $\mathbb{Z}_3^*$ modulo 3^ℓ, in symbols

$$(3.2) \qquad \pi_\ell(R(\mathcal{F}_{j_\ell, k_\ell})) \supset \pi_\ell(\mathbb{Z}_3^*) \qquad \text{for any} \quad \ell \in \mathbb{N}.$$

Now a computation with cardinalities will produce a contradiction in the asymptotics.

$$
\begin{aligned}
1 &\leqslant \frac{|\pi_\ell(R(\mathcal{F}_{j_\ell, k_\ell}))|}{|\pi_\ell(\mathbb{Z}_3^*)|} && \text{by (3.2),}\\[2mm]
&\leqslant \frac{|\mathcal{F}_{j_\ell, k_\ell}|}{|\pi_\ell(\mathbb{Z}_3^*)|} &&\\[2mm]
&= \frac{1}{2 \cdot 3^{\ell-1}} \binom{j_\ell + k_\ell}{j_\ell} && \text{combinatorics (cf. cor. 1.8),}\\[2mm]
&\leqslant \frac{1}{2 \cdot 3^{\ell-1}} \binom{j_\ell + k_\ell}{\lfloor (j_\ell + k_\ell)/2 \rfloor} && \text{binomial coefficients!}\\[2mm]
&\leqslant \frac{c \cdot 2^{j_\ell + k_\ell}}{2 \cdot 3^{\ell-1} \sqrt{j_\ell + k_\ell}} && \text{for some constant } c > 0,\\[2mm]
&= \frac{3c}{2\sqrt{j_\ell + k_\ell}} \cdot 2^{j_\ell + k_\ell - \alpha \ell} && \text{because} \quad \alpha = \log_2 3,\\[2mm]
&\leqslant \frac{3c}{2\sqrt{j_\ell + k_\ell}} \exp\big(\log 2(|j_\ell - \eta \ell| + |k_\ell - \xi \ell|)\big) && \text{because} \quad \xi + \eta = \alpha.
\end{aligned}
$$

But this last product tends to zero for $\ell \to \infty$: the first factor goes to zero as $j_\ell + k_\ell \to \infty$, and the second factor is bounded by condition (iv) of definition 3.1. $\square$

The reduction theorem. The essential point in the proof of theorem 3.5 was condition (iv) of definition 3.1 concerning the speed of convergence of the sequences (j_ℓ/ℓ) and (k_ℓ/ℓ). On the other hand, the proof of lemma 3.2 given here makes use of just this condition, because a condition about the speed of convergence is needed to prove positive asymptotic predecessor density via theorem 2.1. But we *could* prove *sub-positive* asymptotic predecessor density using a *sub-optimal* sequence, provided such a sequence exists. Let us make this more precise.

3.6. DEFINITION. A sequence of pairs $(j_\ell, k_\ell)_{\ell \in \mathbb{N}}$ is called a *globally sub-optimal sequence*, if the following three conditions are satisfied:

(i) For each $\ell \in \mathbb{N}$, the triple (j_ℓ, k_ℓ, ℓ) is globally covering in $\mathbb{Z}_3^*$.
(ii) The limits $\eta := \lim_{\ell \to \infty}(j_\ell/\ell)$ and $\xi := \lim_{\ell \to \infty}(k_\ell/\ell)$ exist.
(iii) These limits fulfill $\eta < 1$ and $\xi + \eta = \alpha$ with $\alpha = \log_2 3$.

3.7. LEMMA. *Suppose there exists a globally sub-optimal sequence. Then*

$$\liminf_{x \to \infty} \left(\inf_{a \in \mathbb{N} \setminus 3\mathbb{N}} \frac{Z_a(ax)}{x^\delta} \right) > 0 \qquad \text{for any } \delta \in \mathbb{R} \text{ satisfying } 0 < \delta < 1 .$$

PROOF. Let $(j_\ell, k_\ell)_{\ell \in \mathbb{N}}$ be a globally sub-optimal sequence. By (i) of the above definition, and by definition 1.12, we know that each triple (j_ℓ, k_ℓ, ℓ) is locally covering at each point $a \in \mathbb{Z}_3^*$. We infer from lemma 1.3 that

$$e_\ell(k, a) \geqslant \binom{k - k_\ell + \ell - j_\ell}{\ell - j_\ell} \qquad \text{for all } k, \ell \in \mathbb{N}.$$

Moreover, by condition (iii) of definition 3.6, we conclude that the limits η and ξ satisfy $\eta < 1$ and $2 - \alpha + \xi - \eta = 2(1 - \eta) > 0$. Now theorem 2.1 applies and proves, for any positive real number $\beta < 2$,

$$\liminf_{n \to \infty} \left(\inf_{a \in \mathbb{Z}_3^*} \frac{s_n(a)}{\beta^n} \right) > 0 .$$

Then theorem III.2.5, taking into account remark III.2.6, shows that

$$\liminf_{x \to \infty} \left(\inf_{a \in \mathbb{N} \cap \mathbb{Z}_3^*} \frac{Z_a(ax)}{x^{\log_2(\beta)}} \right) > 0 .$$

Finally, by the set equations $\mathbb{N} \cap \mathbb{Z}_3^* = \{ a \in \mathbb{N} : a \not\equiv 0 \bmod 3 \} = \mathbb{N} \setminus 3\mathbb{N}$, the proof is complete. □

The critical condition in this lemma is the existence of a globally sub-optimal sequence. We shall see that this existence can be guaranteed provided the sets $\mathcal{R}_{j,k}$ of mixed power sums (notation 1.8) are not too badly distributed. A reasonable conjecture is the following:

3.8. COVERING CONJECTURE FOR MIXED POWER SUMS. *There is a constant $K > 0$ such that, for every $j, \ell \in \mathbb{N}$, the following implication holds:*

$$|\mathcal{R}_{j-1,j}| \geqslant K \cdot 2 \cdot 3^{\ell-1} \quad \Longrightarrow \quad \mathcal{R}_{j-1,j} \text{ covers } \mathbb{Z}_3^* \text{ modulo } 3^\ell .$$

In other words, the meaning of this conjecture is: *If a set $\mathcal{R}_{j-1,j}$ has sufficiently many elements to cover the set of prime residue classes modulo 3^ℓ K times, where K is arbitrarily large, but does not depend neither on j nor on ℓ, then the actual distribution of $\mathcal{R}_{j-1,j}$ among these prime residue classes modulo 3^ℓ is so uniform that at least one element is found in each residue class.* The indices $i := j - 1$ and $k := j$ are chosen to make to the cardinality of $\mathcal{R}_{i,k}$ large; in fact, we know by corollary 1.8 that

$$|\mathcal{R}_{j-1,j}| = \binom{2j}{j}$$

which is the maximal binomial coefficient with upper entry $2j$.

The proof of existence of a globally sub-optimal sequence will also work with a weaker variant of conjecture 3.8.

3.9. WEAK COVERING CONJECTURE FOR MIXED POWER SUMS. *Suppose that* $K(\ell) > 0$ *is a function growing less than an exponential,*

$$\lim_{\ell \to \infty} K(\ell)\, e^{-\gamma \ell} = 0 \qquad \text{for any constant} \quad \gamma > 0\,.$$

Then, for every $j, \ell \in \mathbb{N}$, *the following implication holds:*

$$|\mathcal{R}_{j-1,j}| \geqslant K(\ell) \cdot 3^{\ell} \quad \Longrightarrow \quad \mathcal{R}_{j-1,j} \text{ covers } \mathbb{Z}_3^* \text{ modulo } 3^{\ell}.$$

3.10. REDUCTION THEOREM. *If the weak covering conjecture for mixed power sums holds, then*

$$\liminf_{x \to \infty} \left(\inf_{a \in \mathbb{N} \setminus 3\mathbb{N}} \frac{Z_a(ax)}{x^{\delta}} \right) > 0 \qquad \text{for any } \delta \in \mathbb{R} \text{ satisfying } 0 < \delta < 1\,.$$

PROOF. In view of lemma 3.7 it suffices to show that there is a globally suboptimal sequence. To construct the sequence, we first calculate the expansion factor of the cardinalities of $\mathcal{R}_{j-1,j}$:

$$\frac{|\mathcal{R}_{j-1,j}|}{|\mathcal{R}_{j-2,j-1}|} = \frac{\binom{2j}{j}}{\binom{2j-2}{j-1}} = \frac{2j(2j-1)}{j^2} = 4 \cdot \left(1 - \frac{1}{2j}\right) < 4\,.$$

This implies that, for each $\ell \in \mathbb{N}$, there is at least one $j_\ell \in \mathbb{N}$ such that

$$(3.3) \qquad K(\ell)\, 3^{\ell} \leqslant |\mathcal{R}_{j_\ell - 1, j_\ell}| = \binom{2j_\ell}{j_\ell} \leqslant 4K(\ell)\, 3^{\ell}\,.$$

Now the weak covering conjecture for mixed power sums implies that each set $\mathcal{R}_{j_\ell - 1, j_\ell}$ covers $\mathbb{Z}_3^*$ modulo 3^{ℓ}. As $\mathbb{Z}_3^*$ is a (multiplicative) group, the translates

$$2^{-2j_\ell} \mathcal{R}_{j_\ell - 1, j_\ell}$$

also cover all of $\mathbb{Z}_3^*$ modulo 3^{ℓ} (here the inverse of 2^{2j_ℓ} is taken in the group $\mathbb{Z}_3^*$). Now corollary 1.8 and the implication (d) $\Rightarrow$ (a) of lemma 1.11 show that each triple (j_ℓ, j_ℓ, ℓ) is locally covering at each point $a \in \mathbb{Z}_3^*$. Thus, condition (i) of definition 3.6 is fulfilled.

To ensure requirements (ii) and (iii) of that definition, it suffices to prove the relations

$$(3.4) \qquad \liminf_{\ell \to \infty} \frac{j_\ell}{\ell} \geqslant \log_4 3 \qquad \text{and} \qquad \limsup_{\ell \to \infty} \frac{j_\ell}{\ell} \leqslant \log_4 3\,.$$

It will turn out that these relations are consequences of the well-known asymptotics of binomial coefficients, and the chain of inequalities (3.3). Known facts about binomial coefficients (see lemma III.3.4) combine to the estimates

$$(3.5) \qquad \binom{2j}{j} \geqslant C_1 \frac{2^{2j}}{\sqrt{j}} \qquad \text{with} \quad C_1 = \frac{1}{\sqrt{\pi}} \exp\left(-\frac{1}{6}\right)$$

and

$$(3.6) \qquad \binom{2j}{j} \leqslant C_2 \, \frac{2^{2j}}{\sqrt{j}} \qquad \text{with} \quad C_2 = \frac{1}{\sqrt{\pi}} \, \exp\left(\frac{1}{24}\right) .$$

To show the (easier) lim inf-estimate of (3.4), we combine (3.3) and (3.6) to

$$K(\ell) \, 3^\ell \leqslant \binom{2j_\ell}{j_\ell} \leqslant C_2 \, \frac{4^j}{\sqrt{j}} ,$$

which implies

$$(3.7) \qquad \frac{j_\ell}{\ell} - \log_4 3 \geqslant \frac{1}{\ell} \log_4 \left(\frac{K(\ell) \sqrt{j_\ell}}{C_2} \right) .$$

We have, clearly, $K(\ell) \geqslant \frac{2}{3}$ in the weak covering conjecture for mixed power sums (otherwise $\mathcal{R}_{j-1,j}$ could not cover $\mathbb{Z}_3^*$ modulo 3^ℓ by its cardinality). It is also clear from (3.3) that j_ℓ is large for large ℓ, whence (3.7) proves the lim inf-part of (3.4).

To prove the (a bit harder) lim sup-part of (3.4), note that, by a similar argument as before, (3.3) and (3.5) have the consequence

$$(3.8) \qquad \frac{j_\ell}{\ell} - \log_4 3 \leqslant \frac{1}{\ell} \log_4 \left(\frac{K(\ell) \sqrt{j_\ell}}{C_1} \right) .$$

The remaining point is to prove that the right hand side tends to zero for $\ell \to \infty$. To this end, observe that the immediate estimate $j_\ell \leqslant 2\ell$ for sufficiently large $\ell \in \mathbb{N}$, together with the growth restriction imposed on $K(\ell)$ by the weak covering conjecture for mixed power sums, imply that, for any constant $\gamma > 0$,

$$K(\ell) \, \sqrt{j_\ell} \leqslant e^{\gamma \ell} \sqrt{2\ell} \qquad \text{for sufficiently large} \quad \ell \in \mathbb{N} .$$

Taking $\gamma \to 0$ we conclude that the right hand side of (3.8) vanishes for $\ell \to \infty$. This implies the lim sup-estimate of (3.4), and the proof of the reduction theorem is complete. $\square$

BIBLIOGRAPHY

[AMT]   Albeverio, S., Merlini, D., and Tartini, R., *Una breve introduzione a diffusioni su insiemi frattali e ad alcuni essempi di sistemi dinamici semplici*, Nota di matematica e fisica, Edizioni Cerfim Locarno **3** (1989), 1–39.

[All]   Allouche, J.-P., *Sur la conjecture de "Syracuse-Kakutani-Collatz"*, Sémin. Théor. Nombres, Univ. Bordeaux I, exposé no. 9 (Talence 1979), 15 pages.

[And]   Anderson, S., *Struggling with the 3x+1 problem*, Math. Gaz. **71** (1987), 271–274.

[AL1]   Applegate, D., and Lagarias, J.C., *Density Bounds for the 3x+1 Problem I. Tree-Search Method*, Math. Comp. **65** (1995), 411–426.

[AL2]   Applegate, D., and Lagarias, J.C., *Density Bounds for the 3x+1 Problem II. Krasikov Inequalities*, Math. Comp. **65** (1995), 427–438.

[AL3]   Applegate, D., and Lagarias, J.C., *On the Distribution of 3x + 1 Trees*, Experimental Math. **4** (1995), 193–209.

[Ars]   Arsac, J., *Algorithmes pour verifier la conjecture de Syracuse*, RAIRO, Inf. Théor. Appl. **21** (1987), 3–9.

[Ash]   Ashbacher, C., *Further investigations of wondrous numbers*, J. Recreational Math. **24** (1992), 12–15.

[Ban]   Banerji, R.B., *Some properties of the 3n+1 function,*, Cybernetics and Systems **27** (1996), 473–486.

[BMi]   Belaga, E. and Mignotte, M., *Embedding the 3x+1 conjecture into a 3x+d context*, preprint (1996).

[BMR]   Beltraminelli, S., Merlini, D., and Rusconi, L., *Orbite inverse nel problema del 3n+1*, Nota di matematica e fisica, Edizioni Cerfim Locarno **7** (1994), 325–357.

[BeM]   Berg, L., and Meinardus, G., *Functional equations connected with the Collatz problem*, Results in Math. **25** (1994), 1–12.

[BeM2] Berg, L., and Meinardus, G., *The 3n+1 Collatz Problem and Functional equations*, Rostock Math. Kolloq. **48** (1995), 11–18.

[Ber]   Berndt, B.C., *Ramanujan's Notebooks, Part II*, Springer, New York, 1989.

[Brs]   Bernstein, D.J., *A Non-Iterative 2-adic Statement of the 3N + 1 Conjecture*, Proc. Amer. Math. Soc. **121** (1994), 405–408.

[BrsL] Bernstein, D.J. and Lagarias, J.C., *The 3x + 1 Conjugacy Map*, Can. J. Math. **48** (1996), 1154–1169.

[BłP]   Błazewicz, J., and Pettorossi, A., *Some Properties of Binary Sequences Useful for Proving Collatz's Conjecture*, Foundations of Control Engineering **8** (1983), 53–63.

[BöS]   Böhm, C., and Sontacchi, G., *On the existence of cycles of given length in integer sequences like $x_{n+1} = x_n/2$ if x_n even, and $x_{n+1} = 3x_n + 1$ otherwise*, Atti Accad. Naz. Lincei, VIII Ser., Rend., Cl. Sci. Fis. Mat. Nat. **LXIV** (1978), 260–264.

[Boyd] Boyd, D., *Which Rationals are Ratios of Pisot Sequences?*, Can. Math. Bull. **23** (1985), 343–349.

[Bro]   Brocco, S., *A Note on Mignosi's Generalization of the 3x + 1 Problem*, J. Number Theory **52** (1995), 173–178.

[Bur]   Burckel, S., *Functional equations associated with congruential functions*, Theoretical Computer Science **123** (1994), 397–406.

[BuM]   Buttsworth, R.N., and Matthews, K.R., *On some Markov matrices arising from the generalized Collatz mapping*, Acta Arith. **LV** (1990), 43–57.

[Cad]   Cadogan, C.C., *A note on the 3x+1 problem*, Caribb. J. Math. **3** (1984), 67–72.

[Cha]   Chamberland, M., *A Continuous Extension of the 3x + 1 Problem to the Real Line*, Dynamics of Continuous, Discrete and Impulsive Systems **2** (1996), 495–509.

[Chi]   Chisala, B.P., *Cycles in Collatz sequences*, Publ. Math. Debrecen **45** (1994), 35–39.

[Cla]   Clark, D., *Second order difference equations related to the Collatz 3n+1 conjecture*, J. Difference Equations & Appl. **1** (1995), 73–85.

[CGV]   Cloney, T., Goles, E., and Vichniac, G.Y., *The 3x+1 Problem: A Quasi Cellular Automaton*, Complex Syst. **1** (1987), 349–360.

[Col1]   Collatz, L., *Verzweigungsdiagramme und Hypergraphen*, International Series for Numerical Mathematics, vol. 38, Birkhäuser, 1977.

[Col2]   Collatz, L., *About the motivation of the (3n+1)-problem*, J. Qufu Norm. Univ., Nat. Sci. **3** (1986), 9–11. (chinese)

[Con]   Conway, J.H., *Unpredictable Iterations*, Proc. 1972 Number Theory Conf., University of Colorado, Boulder, Colorado (1972), 49–52.

[Cra]   Crandall, R.E., *On the "3x+1" Problem*, Math. Comp. **32** (October 1978), 1281–1292.

[Dav]   Davison, J.L., *Some Comments on an Iteration Problem*, Proc. 6th Manitoba Conf. Numerical Mathematics (1976), 155–159.

[DGM]   Dolan, J.M., Gilman, A.F., and Manickam, S., *A Generalization of Everett's Result on the Collatz 3x+1 Problem*, Adv. Appl. Math. **8** (1987), 405–409.

[Edg]   Edgar, G.A., *Measure, Topology, and Fractal Geometry*, Springer-Verlag, New York, 1990.

[Eli]   Eliahou, S., *The 3x+1 problem: new lower bounds on nontrivial cycle length*, Discrete Math. **118** (1993), 45–56.

[Eve]   Everett, C.J., *Iteration of the Number-Theoretic Function $f(2n) = n$, $f(2n + 1) = 3n + 2$*, Adv. Math. **25** (1977), 42–45.

[Fel]   Feller, W., *An Introduction to Probability Theory and Its Applications*, 2nd ed., John Wiley & Sons, New York, 1957.

[FMMT]   Feix, M.R., Muriel, A., Merlini, D., and Tartini, R., *The (3x+1)/2 Problem: A Statistical Approach*, in: Stochastic Processes, Physics and Geometry II, Locarno 1991 (Eds. S. Albeverio, U. Cattaneo, D. Merlini) World Scientific (1995), 289–300.

[FMR]   Feix, M.R., Muriel, A., and Rouet, J.L., *Statistical Properties of an Iterated Arithmetic Mapping*, J. Stat. Phys. **76** (1994), 725–739.

[Fil]   Filipponi, P., *On the 3n + 1 Problem: Something Old, Something New*, Rendiconti di Matematica, Serie VII **11** (1991), 85–103.

[FP]   Franco, Z. and Pomerance, C., *On a Conjecture of Crandall Concerning the $qx + 1$ Problem*, Math. Comp. **64** (1995), 1333–1336.

[Gal]   Gale, D., *Mathematical Enterteinments: More Mysteries*, Math. Intelligencer **13** (1991), 54–55.

[Gao]   Gao, G.-G., *On consecutive numbers of the same height in the Collatz problem*, Discrete Math. **112** (1993), 261–267.

[Grd]   Gardner, M., *Mathematical Games*, Scientific American **226** (Juni 1972), 114–118.

[Grn]   Garner, L.E., *On Heights in the Collatz 3n+1 Problem*, Discrete Math. **55** (1985), 57–64.

[GlW]   Glaser, H. und Weigand, H.-G., *Das ULAM-Problem — Computergestützte Entdeckungen*, DdM **2** (1989), 114–134.

[Guy1]   Guy, R.K., *Unsolved Problems in Number Theory*, Springer, New York, 1981, Problem E16.

[Guy2]   Guy, R.K., *Don't try to solve these problems!*, Am. Math. Mon. **90** (1983), 35–41.

[Guy3]   Guy, R.K., *John Isbell's Game of Beanstalk and John Conway's Game of Beans-Don't-Talk*, Math. Mag. **59** (1986), 259–269.

[Has]   Hasse, H., *Vorlesungen über Zahlentheorie*, Zweite Auflage, Springer, Berlin, 1964.

[Hay]   Hayes, B., *Computer Recreations: On the Ups and Downs of Hailstone Numbers*, Scientific American **250** (1984), 10–16.

[Hen]   Hensel, K., *Zahlentheorie*, G. J. Göschen'sche Verlagshandlung, Berlin und Leipzig, 1913.

[Hep]   Heppner, E., *Eine Bemerkung zum Hasse-Syracuse Algorithmus*, Arch. Math. **31** (1978), 317–320.

[Hof]   Hofstadter, D.R., *Gödel, Escher, Bach*, Penguin Books, Harmondsworth, 1980.

[Jar]   Jarvis, F., *13, 31 and the 3x+1 problem*, Eureka **49** (1989), 22–25.

[Jeg]   Jeger, M., *Computer-Streifzüge. Eine Einführung in Zahlentheorie und Kombinatorik aus algorithmischer Sicht*, Birkhäuser, Basel, 1986.

[Kel]   Keller, G., letter to K.P. Hadeler dated 9.4.84 (1984).

[Kor1]   Korec, I., *The 3x+1 problem, generalized Pascal triangles, and cellular automata*, Math. Slovaca **42** (1992), 547–563.

[Kor2]   Korec, I., *A density estimate for the 3x+1 problem*, Math. Slovaca **44** (1994), 85–89.

[KoZ]   Korec, I., and Znám, Š., *A Note on the 3x+1 Problem*, Am. Math. Mon. **94** (1987), 771–772.

[Krf]   Krafft, V., private communication (1992).

[Krs]   Krasikov, I., *How many numbers satisfy the 3x+1 conjecture?*, Int. J. Math. Math. Sci. **12** (1989), 791–796.

[Kut]   Kuttler, J.R., *On the 3x + 1 Problem*, Adv. Appl. Math. **15** (1994), 183–185.

[Lag1]   Lagarias, J.C., *The 3x+1 Problem and its Generalizations*, Am. Math. Mon. **92** (1985), 1–23.

[Lag2]   Lagarias, J.C., *The set of rational cycles for the 3x+1 problem*, Acta Arith. **LVI** (1991), 33–53.

[Lag3]   Lagarias, J.C., *3x + 1 Problem Annotated Bibliography*, september 22, 1997.

[LaW]   Lagarias, J.C., and Weiss, A., *The 3x+1 Problem: Two Stochastic Models*, Ann. Appl. Probab. **2** (1992), 229–261.

[Lea]   Leavens, G.T., *A Distributed Search Program for the 3x+1 Problem*, Technical Report 89-22, Department of Computer Science, Iowa Sate University, Ames, Iowa 50011, USA, 1989.

[LeV]   Leavens, G.T., and Vermeulen, M., *3x+1 Search Programs*, Comput. Math. Appl. **24** (1992), 79–99.

[Lei]   Leigh, G.M., *A Markov process underlying the generalized Syracuse algorithm*, Acta Arithmetica **XLVI** (1986), 125–143.

[LoP]   Lovász, L. and Plummer, M.D., *Matching Theory*, North-Holland, Amsterdam, 1986, pp. 78–81.

[Mah]   Mahler, K., *p-adic Numbers and Their Functions*, Cambridge University Press, Cambridge, 1981.

[Mar]   Marcu, D., *The powers of two and three*, Discrete Math. **89** (1991), 211–212.

[ML]   Matthews, K.R., and Leigh, G.M., *A generalization of the Syracuse algorithm in $F_q[x]$*, J. Number Theory **25** (1987), 274–278.

[MW1]   Matthews, K.R., and Watts, A.M., *A generalization of Hasse's generalization of the Syracuse algorithm*, Acta Arith. **XLVIII** (1984), 167–175.

[MW2]   Matthews, K.R., and Watts, A.M., *A Markov approach to the generalized Syracuse algorithm*, Acta Arith. **XLV** (1985), 29–42.

[Mat1]   Matthews, K.R., *Some Borel measures associated with the generalized Collatz mapping*, Colloq. Math. **63** (1992), 191–202.

[Mat2]   Matthews, K.R., *A survey of some recent work on the generalized Collatz mapping*, preprint (1992).

[Mei]   Meinardus, G., *Über das Syracuse-Problem*, Preprint Nr. 67, Universität Mannheim (1987).

[MSS1]   Merlini, D., Sala, M., and Sala, N., *The 3n+1 Problem and the calculus of the Collatz's constant*, Cerfim, P.O.Box 1132, 6601 Locarno, Switzerland, PC 23/95 (1995).

[MSS1]   Merlini, D., Sala, M., and Sala, N., *On the stopping constant in the 3n+1 problem*, Cerfim, P.O.Box 1132, 6601 Locarno, Switzerland, PC 24/96 (1996).

[Mic]   Michel, P., *Busy beaver competition and Collatz-like problems*, Arch. Math. Logic **32** (1993), 351–367.

[Mig]   Mignosi, F., *On a Generalization of the 3x + 1 problem*, J. Number Theory **55** (1995), 28–45.

[Möl]   Möller, H., *Über Hasses Verallgemeinerung des Syracuse-Algorithmus (Kakutanis Problem)*, Acta Arith. **XXXIV** (1978), 219–226.

[Mül1]   Müller, H., *Das '3n+1' Problem*, Mitt. Math. Ges. Hamb. **12** (1991), 231–251.

[Mül2]   Müller, H., *Über eine Klasse 2-adischer Funktionen im Zusammenhang mit dem "3x + 1"-Problem*, Abh. Math. Sem. Univ. Hamburg **64** (1994), 293–302.

[NFR]   Nievergelt, J., Farrar, J.C., and Reingold, E.M., *Computer approaches to mathematical problems*, Prentice Hall, Inc., Englewood cliffs, N. J., 1974.

[Ogi]   Ogilvy, C.S., *Tomorrow's math*, 2nd ed., Oxford University Press, London, 1972.

[Pen]   Penning, P., Crux Math. **15** (1989), 282–283.

[Pic]   Pickover, C.A., *Hailstone (3n+1) Number Graphs*, J. Recreational Math. **21** (1989), 120–123.

[Pud]   Puddu, S., *The Syracuse problem*, Notas Soc. Mat. Chile **5** (1986), 199–200, see MR 88c:11010. (spanish)

[Raw]   Rawsthorne, D.A., *Imitiation of an Iteration*, Math. Mag. **58** (1985), 172–174.

[Rev]   Revuz, D., *Markov Chains*, North-Holland, Amsterdam, 1984.

[Ro1]   Rozier, O., *Démonstration del l'absence de cycles d'une certain forme pour le problème de Syracuse*, Singularité 1 (1990), 9–12.

[Ro2]   Rozier, O., *Problème de Syracuse: "Majorations" Elementaires Diophantiennes et Nombres Transcendent*, Singularité 2 (1991), 8–11.

[San]   Sander, J.W., *On the (3N+1)-Conjecture*, Acta Arith. **LV** (1990), 241–248, the result already had been presented during a conference in 1987.

[Sch]   Schuppar, B., *Kettenbrüche und der (3n+1)-Algorithmus*, unveröffentlicht (1981).

[Sei]   Seifert, B.G., *On the Arithmetic of Cycles for the Collatz-Hasse ('Syracuse') Conjecture*, Discrete Math. **68** (1988), 293–298.

[Sha]   Shallit, J.O., *The "3x + 1" Problem and Finite Automata*, Bull. EATCS (European Assoc. for Theor. Comp. Sci.) **46** (1991), 182–185.

[Ste1]   Steiner, R.P., *A Theorem on the Syracuse Problem*, Proc. 7th Manitoba Conf. Numerical Mathematics and Computing 1977 Winnipeg (1978), 553–559.

[Ste2]   Steiner, R.P., *On the "Qx+1 Problem," Q odd*, Fibonacci Q. **19** (1981), 285–288.

[Ste3]   Steiner, R.P., *On the "Qx+1 Problem," Q odd, II*, Fibonacci Q. **19** (1981), 293–296.

[Tar]   Targonski, Gy., *Open questions about KW-orbits and iterative roots*, Aequationes Math. **41** (1991), 277–278.

[Ter1]   Terras, R., *A stopping time problem on the positive integers*, Acta Arith. **XXX** (1976), 241–252.

[Ter2]   Terras, R., *On the existence of a density*, Acta Arith. **XXXV** (1979), 101–102.

[Thw]   Thwaites, B., *My Conjecture*, Bull., Inst. Math. Appl. **21** (1985), 35–41.

[Ven1]   Venturini, G., *Sul Comportamento delle Iteratzione di Alcuni Funzioni Numeriche*, Rend. Sci. Math. Institute Lombardo **A 116** (1982), 115–130.

[Ven2]   Venturini, G., *On the 3x+1 Problem*, Adv. Appl. Math. **10** (1989), 344–347.

[Ven3]   Venturini, G., *Iterates of number theoretic functions with periodic rational coefficients (generalization of the 3x+1 problem)*, Studies in Appl. Math. **86** (1992), 185–218.

[Ven4]   Venturini, G., *On a generalization of the 3x+1 Problem*, Adv. Appl. Math. **19** (1997), 332–354.

[Wag]   Wagon, S., *The Collatz Problem*, Math. Intelligencer **7** (1985), 72–76.

[Wig]   Wiggin, B.C., *Wondrous Numbers — Conjecture abouts the 3n + 1 Family*, J. Recreational Math. **20** (1988), 52–56.

[W....]   Williams, C., Thwaites, B., van der Poorten, A., Edwards, W., and Williams, L.,
          Ulam's conjecture continued again, PPC Calculator Journal **9** (September 1982),
          23–24.

[Wls]   Wilson, D.W., *A link from the Collatz conjecture to finite automata*, distributed
          via email (1991), dwilson@cvbnet.prime.com.

[Wir1]   Wirsching, G.J., *An improved estimate concerning 3n+1 predecessor sets*, Acta
          Arith. **LXIII** (1993), 205–210.

[Wir2]   Wirsching, G.J., *On the combinatorial structure of 3n+1 predecessor sets*, Discrete
          Math. **148** (1996), 265–286.

[Wir3]   Wirsching, G.J., *A Markov chain underlying the backward Syracuse algorithm*,
          Rev. Roum. Math. Pures Appl. **XXXIX** (1994), 915–926.

[Wir4]   Wirsching, G.J., *3n+1 predecessor densities and uniform distribution in* $\mathbb{Z}_3^*$, Proc.
          conf. Elementary and Analytic Number Theory, Vienna, july 18–20, 1996, Eds. W.
          G. Nowak and J. Schoißengeier (1997), 230–240.

[Wu1]   Wu, J.B., *The trend of changes of the proportion of consecutive numbers of the
          same height in the Collatz problem*, J., Huazhong (Central China) Univ. Sci. Tech-
          nol., **20** (1992), 171–174, MR 94b:11024. (chinese)

[Wu2]   Wu, J.B., *On the consecutive positive integers of the same height in the Collatz
          problem*, Math. Appl., suppl. **6** (1993), 150–153. (chinese)

[Yam]   Yamada, M., *A Convergence proof About an Integral Sequence*, Fibonacci Q. **18**
          (1980), 231–242, See MR 82d:10026 for a serious flaw in the proof.

INDEX OF AUTHORS

LIST OF SYMBOLS

Greek characters.

$\Gamma = (V, E)$	directed graph 33.
$\widetilde{\Gamma}(a)$	weak component of Γ containing the vertex a 34.
$\Gamma_f = (V_f, E_f)$	Collatz graph of f 36.
$\Gamma_T^* = (V_T^*, E_T^*)$	pruned Collatz graph 58.
$\Gamma_\ell = 3^\ell$	cardinality of K_ℓ 100.
$\Delta_{\ell+1}(k', a')$	domain of dependence 109.
$\widetilde{\Delta}_{\ell+1}(k', a')$	lifted domain of dependence 111.
$\Delta_\ell^*(k, a)$	domain of transition 109.
$\widetilde{\Delta}_\ell^*(k, a)$	lifted domain of transition 111.
$\lambda = \log_2(\frac{3}{2})$	a constant 54.
λ_3	natural measure on 3-adic fractions 104.
$\lambda_k(n)$	k-th forward coefficient of n 17.
$\Lambda_\ell = 2 \cdot 3^{\ell-1}$	number of prime residues modulo 3^ℓ 100.
μ_3	normalized Haar measure on $\mathbb{Z}_3$ 79.
ν_3	normalized Haar measure on $\mathbb{Z}_3^*$ 79.
ξ_ℓ	order-preserving projection of 3-adic fractions 106.
$\pi^C(x)$	Crandall's counting function 67.
$\pi_h^C(x)$	Crandall's counting function for fixed height 67.
π_ℓ	natural projection $\mathbb{Z}_3 \twoheadrightarrow \mathbb{Z}_3/3^\ell\mathbb{Z}_3$ 124.
$\Pi(\Gamma)$	set of all paths of the graph Γ 33.
$\varrho = \lambda_3 \otimes \nu_3$	reference measure on state space 106.
$\varrho_k(n)$	k-th forward remainder of n 17.
σ	map $\Pi(\Gamma_T) \to \mathcal{F}$ encoding paths 39.
$\sigma(n)$	stopping time 12,19.
$\tau(n)$	coefficient stopping time 19.
$\tau^s(M)$	transition measure of a Markov chain 108.
Φ	Bernstein's inverse of $Q_0 = Q_\infty$ 26.
$\widetilde{\varphi}_\ell$	pull-back of counting function 106.
χ	Terras' notation for stopping time 19.
$\Omega_f(a)$	cycle generated by a 10.

Latin characters.

$\mathcal{A}_f(a)$	domain of attraction 36.
$\mathbb{B}_3$	set of 3-adic fractions 104.
$B_a(n)$	Crandall's one-step predecessor 65.

INDEX

Druck: Strauss Offsetdruck, Mörlenbach
Verarbeitung: Schäffer, Grünstadt